Mathematics for

Book 2

Morag McClurg and Alan Caldow

Cambridge University Press

Cambridge
New York   Port Chester
Melbourne   Sydney

Published by the Press Syndicate of the University of Cambridge
The Pitt Building, Trumpington Street, Cambridge CB2 1RP
40 West 20th Street, New York, NY 10011, USA
10 Stamford Road, Oakleigh, Melbourne 3166, Australia

© Cambridge University Press 1990

First published 1990

Printed in Great Britain by Scotprint Ltd., Musselburgh

British Library cataloguing in publication data

ISBN 0 521 36902 9

Diagrams and typesetting by DMD Ltd, St Clements, Oxford

Contents

1	Difference of two squares	1
	Money management 6: Mortgages	9
2	Factors of quadratics	13
3	Trigonometric graphs	24
	Consolidation 1	32
4	Indices	35
5	Graphs 2	44
6	Similar areas and volumes	49
	Money management 7: Energy bills	64
7	Circles	68
8	Algebraic fractions	84
	Consolidation 2	95
9	The sine rule and the area of a triangle	99
10	Quadratic equations	110
	Money management 8: Time for a holiday!	127
11	Linear programming	133
12	Graphs 3	146
	Consolidation 3	162
	General review	170

1 Difference of two squares

A A square within a square

Draw a square of side 10 cm on a sheet of paper. Inside it draw a square of side 4 cm, like this.

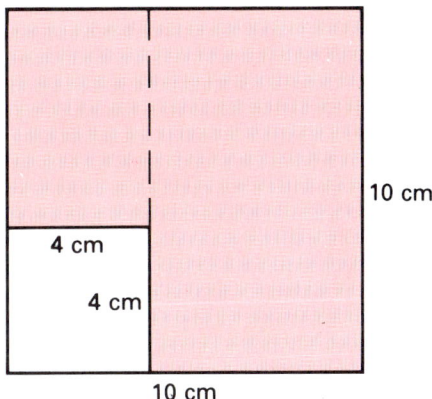

A1 Calculate the area of the shaded part of the square.

Cut the small square out of the large square, and throw it away.
Cut the 'L' shape along the dashed line.

A2 What shapes do you have now? What are their dimensions?

Fit the two shapes together to make a large rectangle.

A3 What is the length and breadth of the large rectangle? Calculate its area.

You now have calculated the shaded area in two ways.

A4 Copy and complete:

$10^2 - 4^2 = \ldots \times \ldots$

$10^2 - 4^2 = (10 - 4)(10 + 4)$

A5 (a) Do the same, starting with a square of side 8 cm and cutting a square of side 3 cm from it.

(b) Write an equation like the one in question A4 for these rectangles.

B The general case

Suppose you have a square of side x cm and cut a square of side y cm from it, like this.

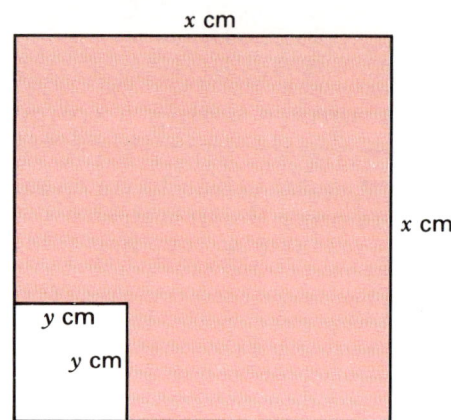

B1 Copy and complete:

The shaded area (in cm²) = $x^2 - \ldots$

Copy the diagram. Draw in a dashed line to split the L-shape into two rectangles.

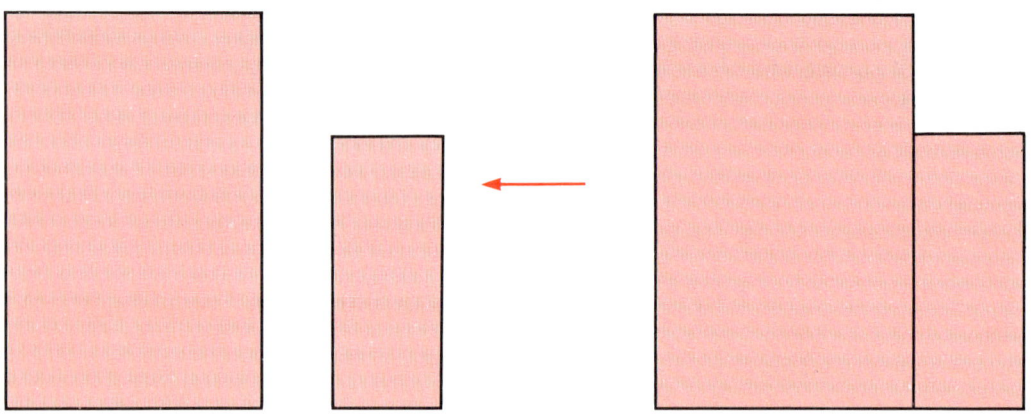

B2 Copy and complete:

Length of small rectangle = ... cm
Width of small rectangle = ... cm
Length of large rectangle = ... cm
Width of large rectangle = ... cm

Since the length of the small rectangle equals the width of the large rectangle, they will fit together like this.

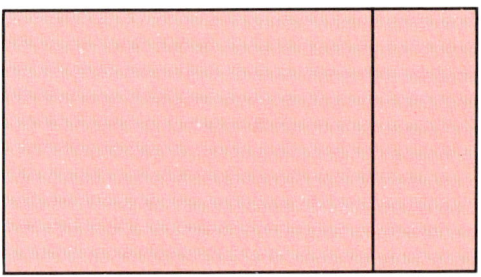

B3 Copy and complete:

The length of the final rectangle = ...
The width of the final rectangle = ...

B4 Write down an expression for the area of the final rectangle.

B5 Copy and complete: $x^2 - y^2 = ...$

C Using the rule

You should now realise that:

$x^2 - y^2 = (x - y)(x + y)$
$x^2 - y^2 = (x + y)(x - y)$

These are both **factorised** forms of $x^2 - y^2$.

C1 Why does it not matter if the order of the brackets is changed?

C2 By multiplying out the brackets, check that $(x + y)(x - y)$ and $(x - y)(x + y)$ both equal $x^2 - y^2$.

C3 Why is this rule called the **difference** of two squares?

It can be easier to remember the rule using words, like this.

Red square minus black squared equals red minus black times red plus black.

Here are some examples.

(1) $x^2 - 16 = x^2 - 4^2 = (x - 4)(x + 4)$

Check: $(x - 4)(x + 4) = x^2 + 4x - 4x - 4^2 = x^2 - 16$

(2) $p^2 - 9 = p^2 - 3^2 = (p - 3)(p + 3)$

Check: $(p - 3)(p + 3) = p^2 + 3p - 3p - 3^2 = p^2 - 9$

(3) $9c^2 - d^2 = (3c)^2 - d^2 = (3c - d)(3c + d)$

Check: $(3c - d)(3c + d) = 9c^2 + 3cd - 3cd - d^2 = 9c^2 - d^2$

(4) $16a^2 - 49e^2f^2 = (4a)^2 - (7ef)^2 = (4a - 7ef)(4a + 7ef)$

Check: $(4a - 7ef)(4a + 7ef) = 16a^2 + 28ef - 28ef - 49e^2f^2$
$= 16a^2 - 49e^2f^2$

C4 Factorise these differences of two squares. Check your answers by multiplying out the brackets.

(a) $b^2 - 7^2$ (b) $6^2 - h^2$ (c) $m^2 - n^2$

(d) $p^2 - q^2$ (e) $x^2 - 25$ (f) $16 - y^2$

(g) $a^2 - 49$ (h) $4d^2 - 1$ (i) $36 - 25f^2$

(j) $9g^2 - 64$ (k) $81 - 16h^2$ (l) $25v^2 - w^2$

(m) $4a^2 - b^2$ (n) $49k^2 - 41^2$ (o) $16d^2 - 9r^2$

(p) $121p^2 - 36q^2$

C5 Now try these.

(a) $4a^2b^2 - 9$ (b) $16 - 9p^2q^2$ (c) $25e^2 - x^2y^2$

(d) $64c^2 - 9g^2h^2$ (e) $36m^2n^2 - 25x^2$ (f) $a^4 - 9$

(g) $64 - 9x^4$ (h) $49x^4 - 9a^2b^2$ (i) $9y^2z^4 - 64$

(j) $d^4 - 16$ (k) $16x^4 - 1$ (l) $81b^4 - 256a^4$

(m) $36d^4 - g^4$ (n) $a^4b^4 - 81$ (o) $625x^4 - y^4z^8$

(p) $16a^4x^4 - 81y^8$

You can use this kind of factorisation to simplify calculations like $51^2 - 49^2$. It turns a subtraction into a multiplication.

C6 Copy and complete:

$51^2 - 49^2 = (51 - \ldots)(\ldots + \ldots)$
$= 2 \times \ldots$
$= \ldots$

C7 Now try these.

(a) $11^2 - 9^2$ (b) $57^2 - 23^2$ (c) $189^2 - 11^2$
(d) $7\cdot77^2 - 2\cdot23^2$ (e) $89\cdot7^2 - 10\cdot3^2$ (f) $3\cdot25^2 - 5\cdot75^2$

D Using two types of factorisation

In SMP 11–16 *Book Y2*, we met the common factor type of factorisation, like

$4a - 28 = 4(a - 7)$

This section deals with expressions which are a combination of a common factor and a difference of two squares.

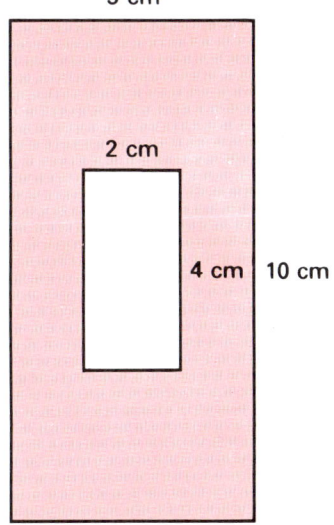

Ann is designing a letter 'O' for a poster. She starts with a rectangle 5 cm by 10 cm and cuts a rectangle 2 cm by 4 cm out of the middle of it.

The area of the 'O' is

$50 \text{ cm}^2 - 8 \text{ cm}^2 = 42 \text{ cm}^2$

We could also work out the area by halving the shape like this.

The area of the 'O' can now be calculated using

$$\begin{aligned}2 \times (25 - 4) \text{ cm}^2 &= 2 \times (5^2 - 2^2) \text{ cm}^2 \\ &= 2 \times (5 - 2)(5 + 2) \text{ cm}^2 \\ &= 2 \times 3 \times 7 \text{ cm}^2 \\ &= 42 \text{ cm}^2\end{aligned}$$

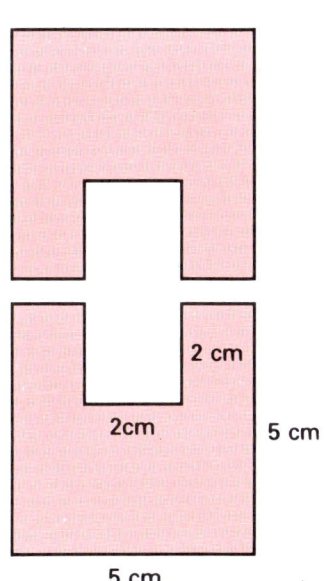

The two different ways of calculating the area of the 'O' are equal, so

$50 - 8 = 2 \times (5^2 - 2^2)$

This illustrates how a difference like $50 - 8$ can be turned into a difference of two squares by taking out a common factor.

Here is an example.

D1 Why is it not possible to use the difference of two squares with this expression?

Here is the solution.

$3x^2 - 48 = 3(x^2 - 16)$
$ = 3(x^2 - 4^2)$
$ = 3(x - 4)(x + 4)$

Common factor of 3

Difference of two squares

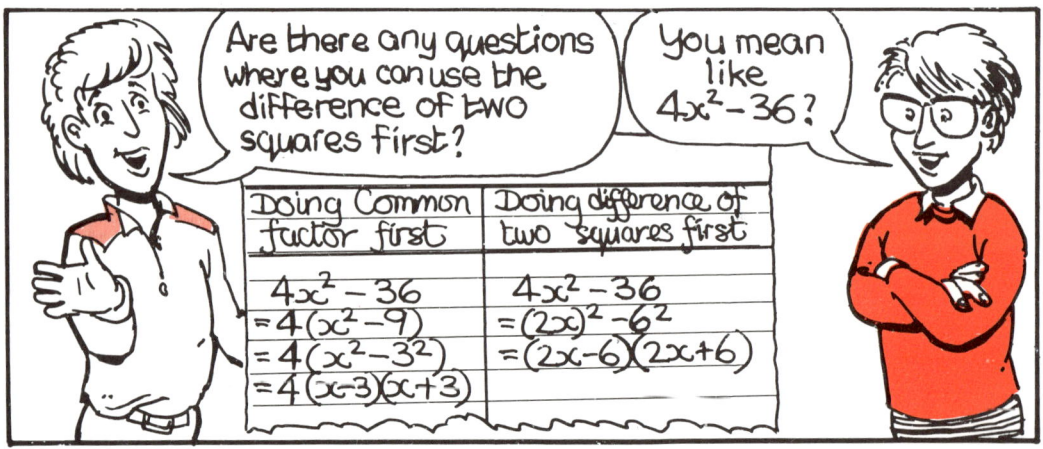

D2 Factorise these.

(a) $3a^2 - 27$
(b) $4p^2 - 4$
(c) $8c^2 - 50$
(d) $12x^2 - 48$
(e) $6 - 54w^2$
(f) $ax^2 - 9a$
(g) $x^2 - x^4$
(h) $4x^3 - 9x$
(i) $7 - 175a^2$
(j) $a - 4ab^2$
(k) $48p^2 - 75q^2$
(l) $3p - 12p^3$
(m) $4ab^2 - 36a^3$
(n) $x^3y - xy^5$
(o) $12a^3b - 27ab^5$
(p) $3a^4b^2 - 27a^2b^4$

E Expressions involving brackets

Here is another difference of two squares:
$(x + 2)^2 - 4$

> Remember the rule
> red^2 − black^2 = (red − black)(red + black)

Using the rule we have

$(x + 2)^2 - 2^2 = ((x + 2) - 2)((x + 2) + 2)$
$= (x + 2 - 2)(x + 2 + 2)$
$= (x)(x + 4)$
$= x(x + 4)$

> Now remove the inside brackets.

Here is another example.

$(x + y)^2 - (x - y)^2$
$= (x + y)^2 - (x - y)^2 = ((x + y) - (x - y))((x + y) + (x - y))$
$= (x + y - x + y)(x + y + x - y)$
$= (2y)(2x)$
$= 4xy$

> Removing the inside brackets.

E1 Factorise these.

(a) $16 - (x - y)^2$
(b) $(x + 4)^2 - 25$
(c) $100 - (a + 3)^2$
(d) $(2x + 1)^2 - 4$
(e) $(2x - y)^2 - (x - 2y)^2$
(f) $(a + b)^2 - (b - c)^2$
(g) $(4x + 1)^2 - (x - 3)^2$
(h) $1 - 16(x - 1)^2$
(i) $(a - b)^2 - 9(a + b)^2$
(j) $4(x + p)^2 - (x - 2p)^2$
(k) $4(a + d)^2 - 9(a - d)^2$
(l) $16(x + 3y)^2 - 25(4y + x)^2$

E2 Here is a mixture of expressions to factorise.

(a) $(a + b - c)^2 - p^2$
(b) $x^2 - (x + y - z)^2$
(c) $12 - 27(a + 1)^2$
(d) $16a^4 - b^4$
(e) $2ab^4 - 32a^5$
(f) $2(x + y)^2 - 8(x - y)^2$
(g) $(a^2 - ab)^2 - (ab - b^2)^2$
(h) $(p + q + r)^2 - (p - q + r)^2$
(i) $3x^4 - 12(x - 4)^2$
(j) $25x^2 - \frac{1}{4}y^2$
(k) $9(a + 2b)^2 - 16(a - b)^2$
(l) $3(x + 1) - 27(x + 1)^3$
(m) $2a^3b(c + 4)^2 - 8ab(1 - c)^2$
(n) $x^8 - 1$

E3 Without evaluating the squares, find by how much $99 \cdot 8^2$ exceeds $99 \cdot 2^2$.

Money management 6: Mortgages

A Mortgages

A mortgage is a loan taken out to buy a house. It is taken out for a long term – usually 20 or 25 years – for a relatively large sum of money. Before granting a mortgage, the bank or building society look at the applicant's salary, and the value of the property. They will lend **either** a multiple of the applicant's salary **or** a percentage of the value of the house, whichever of these is **lower**. Here are the lending policies of three building societies.

H = higher income L = lower income S = single income

Name of society	I Double income	Single income	V Value of property
Rightway	$2\frac{1}{4}H + 1\frac{1}{4}L$	$2\frac{1}{2}S$	
Kirk International	$2\frac{1}{2}H + 1\frac{1}{4}L$	$2\frac{3}{4}S$	100%
Easymoney	$2\frac{1}{2}H + L$	$2\frac{5}{8}S$	95%
In each case the maximum mortgage is the lower of I and V			

Marie and Jim are considering buying this property, which costs £32 000. Jim earns £10 500 and Marie £22 155.

They apply to the Rightway Building Society. Here is part of the reply.

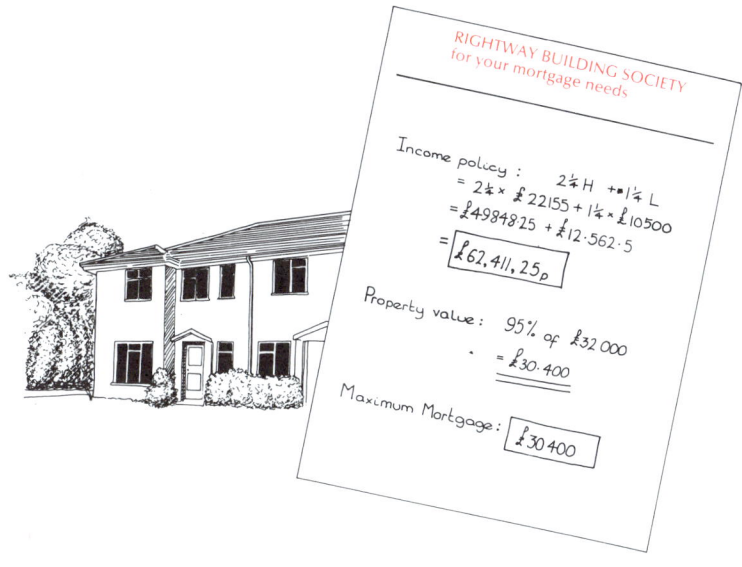

So if they wanted to buy the property they would have to pay a deposit of £1600.

A1 Cathy earns £10 025 and Colin £11 200. Work out the maximum mortgage that each of the above societies would grant, based on their salaries.

A2 Cathy and Colin have been looking at the following properties.

What would be the maximum mortgage each society would give them to purchase each house?

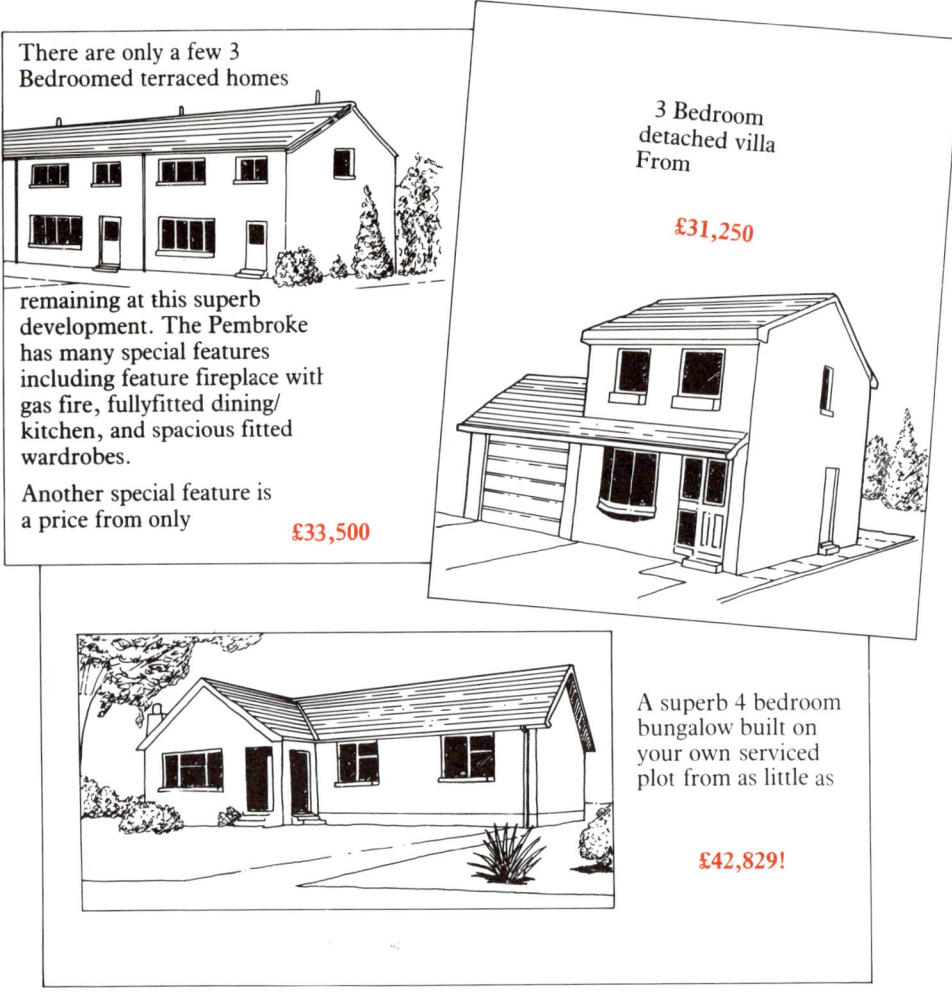

There are only a few 3 Bedroomed terraced homes remaining at this superb development. The Pembroke has many special features including feature fireplace with gas fire, fullyfitted dining/ kitchen, and spacious fitted wardrobes.

Another special feature is a price from only £33,500

3 Bedroom detached villa From

£31,250

A superb 4 bedroom bungalow built on your own serviced plot from as little as

£42,829!

A3 Explain clearly how a single person could get a maximum mortgage of £30 000 from each of the above societies. Remember that you will have to look at both the income policy of each society and the value of the property.

A4 Repeat question A3 for a couple with two incomes.

B Repayments

Like any other loan, compound interest is payable on the amount borrowed. Because of the long term of the loan, working out the repayments can be a complicated process.

Most building societies provide a table like this, as a guide to the cost of repayments.

RIGHTWAY BUILDING SOCIETY		
MONTHLY REPAYMENT GUIDE		
MORTGAGE	PERIOD	MONTHLY PAYMENT
£15 000	25 years	£140·41
£15 000	20 years	£153·49
£15 000	15 years	£177·23
£20 000	25 years	£187·21
£20 000	20 years	£204·65
£20 000	15 years	£236·30
£25 000	25 years	£234·01
£25 000	20 years	£255·81
£25 000	15 years	£295·36
£30 000	25 years	£280·81
£30 000	20 years	£306·99
£30 000	15 years	£354·44
£40 000	25 years	£397·70
£40 000	20 years	£429·50
£40 000	15 years	£489·36
£50 000	25 years	£510·71
£50 000	20 years	£549·48
£50 000	15 years	£623·29

B1 Cathy and Colin decide to buy this house. They put down a deposit of £2000, and take out a mortgage for the rest with the society above.

(a) If they take out a mortgage for £25 000 over 20 years, how much will they pay altogether?

(b) How much interest will they pay?

(c) If they took out the same mortgage over 15 years, how much less interest will they pay?

B2 John and Alison have saved up to buy this flat in Milngavie. They can afford a deposit of £700 and take out a mortgage for the remaining £15 000.

(a) If they repay the mortgage over 15 years, how much interest will they pay?

(b) How much more interest will they pay if they repay the loan over 25 years?

B3 Mr and Mrs Kydd have taken out a mortgage for £40 000 to buy this desirable residence.

If they will repay £113 910 altogether, over what period did they take out the loan?

2 Factors of quadratics

A Where the coefficient of x^2 is 1

In SMP 11–16 *Book Y1*, we learnt how to multiply out brackets. For example, $(x + 2)(x + 3)$ gives $x^2 + 5x + 6$.

Of course, if we multiply out the brackets in $(x + 3)(x + 2)$, we would still get $x^2 + 5x + 6$ because the order of multiplying is unimportant.

In this chapter, we are going to start with a quadratic expression, like $x^2 + 5x + 6$, and find a product of two brackets equal to it, like $(x + 2)(x + 3)$ or $(x + 3)(x + 2)$. This is called **factorising**. We first met this idea in *Book Y2*.

The table for multiplying out $(x + 2)(x + 3)$ is

	x	3
x	x^2	$3x$
2	$2x$	6

So $(x + 2)(x + 3) = x^2 + 3x + 2x + 6$
$= x^2 + 5x + 6$

A1 Make tables for multiplying out these brackets. Write the expressions as in the above example.

(a) $(x + 1)(x + 7)$ (b) $(x - 2)(x + 5)$
(c) $(x + 2)(x - 6)$ (d) $(x - 2)(x - 8)$

A2 Find $*$ in each of these tables.

(a)
	x	2
x	x^2	$*$
5	$5x$	10

(b)
	x	-4
x	x^2	$-4x$
$*$	$7x$	-28

(c)
	x	6
x	x^2	$6x$
$*$		42

(d)
	x	$*$
x	x^2	
2	$2x$	-10

(e)
	x	$*$
x	x^2	-4
		12

A3 There are many values that $*$ and $\blacktriangle$ can take here.

	x	$*$
x	x^2	
$\blacktriangle$		-6

Copy and complete this list of them. The possible combinations are 6 and -1, 1 and -6 . . .

A4 Find the possible values of $*$ and $\blacktriangle$ in these tables.

(a)

	x	$*$
x	x^2	
$\blacktriangle$		5

(b)

	x	$*$
x	x^2	
$\blacktriangle$		-14

(c)

	x	$*$
x	x^2	
$\blacktriangle$		20

(d)

	x	$*$
x	x^2	
$\blacktriangle$		16

(e)

	x	$*$
x	x^2	
$\blacktriangle$		-12

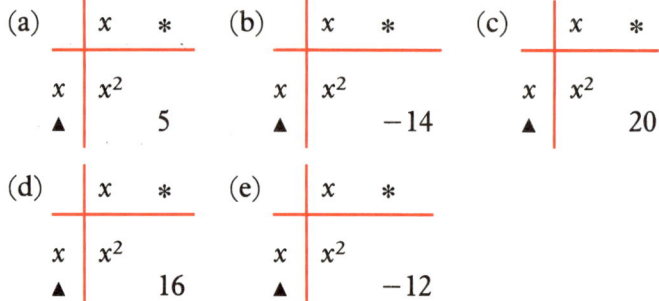

A5 What values must $*$ and $\blacktriangle$ have in this table?

	x	$*$
x	x^2	
$\blacktriangle$		12

These two terms add up to $7x$.

One way to deal with questions like A5 is to make a table showing all the possible values.

$*$	12	-12	6	-6	4	-4
$\blacktriangle$	1	-1	2	-2	3	-3
Adding:	13	-13	8	-8	**7**	-7

All these pairs of numbers multiply to give 12.

This is the combination we are looking for (to give $7x$).
So the values of $*$ and $\blacktriangle$ must be 4 and 3.

The table looks like

	x	3
x	x^2	$3x$
4	$4x$	12

or

	x	4
x	x^2	$4x$
3	$3x$	12

So $x^2 + 7x + 12 = (x + 4)(x + 3)$ or $(x + 3)(x + 4)$.

To factorise an expression like $x^2 + 17x + 30$, the table looks like this.

	x	$*$
x	x^2	
$\blacktriangle$		30

These two terms add to give $17x$

This is the product of $*$ and $\blacktriangle$.

Here is the table of combinations of $*$ and $\blacktriangle$.

$*$	30	-30	15	-15	10	-10	6	-6
$\blacktriangle$	1	-1	2	-2	3	-3	5	-5
Adding:	31	-31	**17**	-17	13	-13	11	-11

All these pairs of numbers multiply to give 30.

The numbers we are looking for are 15 and 2.
So $x^2 + 17x + 30 = (x + 2)(x + 15)$

> We put in the + sign since the numbers are positive.

Of course, we can check this by multiplying out the brackets.

Check

	x	15
x	x^2	$15x$
2	$2x$	30

So $(x + 2)(x + 15) = x^2 + 15x + 2x + 30$
$= x^2 + 17x + 30$

Here is another example. Factorise $x^2 - 2x - 24$.

The table is

	x	*
x	x^2	
▲		-24

> These two terms add to give $-2x$.

Here is the table of combinations of * and ▲.

*	24	-24	12	-12	8	-8	6	-6
▲	-1	1	-2	2	-3	3	-4	4

> All these pairs of numbers multiply to give -24.

Adding: 23 -23 10 -10 5 -5 2 $\boxed{-2}$

The numbers we are looking for are -6 and 4.

So $x^2 - 2x - 24 = (x + 4)(x - 6)$

> The − sign since −6 is negative.
> The + sign since 4 is positive.

Check

	x	-6
x	x^2	$-6x$
4	$4x$	-24

So $(x + 4)(x - 6) = x^2 - 6x + 4x - 24$
$= x^2 - 2x - 24$

A6 Factorise $x^2 + 8x + 12$ like this.

(a) Make a table.

(b) Copy and complete:

*	12	-12	6	-6		
▲	1	-1			3	-3

Adding: 13 -13

The numbers we are looking for are . . . and . . .

So $x^2 + 8x + 12 = (x \ldots)(x \ldots)$

(c) Check your answer.

Factorise these quadratic expressions in the same way.

A7 (a) $x^2 + 16x + 15$ (b) $x^2 + 11x + 30$ (c) $x^2 + 11x + 24$
(d) $x^2 - 6x + 8$ (e) $x^2 + 15x + 44$ (f) $x^2 - 9x + 18$
(g) $x^2 - x - 6$ (h) $x^2 - 2x - 8$ (i) $x^2 + 2x - 3$
(j) $x^2 + 4x - 12$ (k) $x^2 + 3x - 10$ (l) $x^2 + 3x - 4$
(m) $x^2 - 6x - 16$ (n) $x^2 + 16x + 48$ (o) $x^2 - 19x + 60$
(p) $a^2 + 2a - 8$ (q) $b^2 + 7b - 18$ (r) $c^2 + 28c + 75$
(s) $x^2 - 19x + 48$ (t) $d^2 - 11d - 12$ (u) $x^2 + 7x - 8$
(v) $e^2 - e - 20$ (w) $f^2 - 20f + 36$ (x) $x^2 + 16x + 28$

B More tables

B1 Find ∗ and ▲ for these tables.

(a)

	2x	3
▲ 1	$6x^2$ $2x$	$9x$ 3

(b)

	∗	−2
2x 7	$10x^2$	-14

(c)

	3x	1
▲ −3	$6x^2$ $-9x$	

(d)

	∗	1
4x −3	$8x^2$	

(e)

	∗	−3
4x −5	$-5x$	$-12x$ 15

(f)

	3x	4
▲ −1	$-3x$	$8x$ -4

(g)

	∗	−1
▲ 3	$8x^2$	$-x$ -3

(h)

	∗	−2
▲ 4	$9x^2$	$-18x$

(i)

	∗	−1
▲ 3	$8x^2$	$-x$ -3

(j)

	∗	−1
▲ 3	$8x^2$	$-4x$ -3

(k)

	∗	−1
▲ 3	$8x^2$ $3x$	-3

B2 (a) Find $*$ and $\blacktriangle$ in this table.

	$*$	2
$\blacktriangle$	$2x^2$	
1	x	2

(b) If the table were like this, what difference would it make to the possible values for $*$ and $\blacktriangle$?

	$*$	2
$\blacktriangle$	$3x^2$	
1		2

(c) Draw and complete any possible tables for part (b).

The last example shows that there may be more than one possibility for a table. For example, in the table

	$*$	-1
$\blacktriangle$	$5x^2$	
3		-3

$* \times \blacktriangle = 5x^2$ and so one will be $5x$ and the other will be x.

Since we could place these in the table in two different ways the two possibilities for tables are

	$5x$	-1
x	$5x^2$	
3		-3

and

	x	-1
$5x$	$5x^2$	
3		-3

If the tables are completed, they become

	$5x$	-1
x	$5x^2$	$-x$
3	$15x$	-3

so $(x + 3)(5x - 1) = 5x^2 + 14x - 3$

and

	x	-1
$5x$	$5x^2$	$-5x$
3	$3x$	-3

so $(5x + 3)(x - 1) = 5x^2 - 2x - 3$

The only difference in these two expressions is their middle terms (the terms in x). So it matters where the $5x$ and the x go in the table.

17

B3 Find the four different tables possible for each of these. Write each expression.

(a)
*	2
$6x^2$	
	2

▲ 1

(b)
*	2
$8x^2$	
	-6

▲ -3

C Where the coefficient of x^2 is not 1

Consider $(2x + 1)(x + 7)$.
The table for multiplying out the brackets is

	x	7
$2x$	$2x^2$	$14x$
1	x	7

So $(2x + 1)(x + 7) = 2x^2 + 14x + x + 7$
$= 2x^2 + 15x + 7$

The table shows that the numbers in the brackets (the 7 and the 1) still multiply to give the number on its own (7) in the expression.

For the middle term,

$15x = 14x + x = 7 \times 2x + 1 \times x$
$15 = 7 \times 2 + 1 \times 1$

Now consider $(x - 2)(4x - 3) = 4x^2 - 11x + 6$

These multiply to give the 6.

C1 Draw the table for $(x - 2)(4x - 3)$ and use it to explain how to get the middle term in the expression.

Here it is not just a case of finding numbers that add to give something. We have to multiply the number by the coefficient of the x term in the **opposite bracket**, then add.

We can adapt the table of combinations that we used before to help find the correct numbers.

Example 1   Factorise $3x^2 + 7x + 2$.   *(Take the factors of this 3.)*

3	2	1	−2	−1
1	1	2	−1	−2

All these pairs of numbers multiply to give 2.

Note that we have to put **all** the factors of 2 **both ways round** since it matters where the $3x$ and x go in the multiplying-out table.

We can now 'cross-multiply' and add, hoping to get 7.

3	2	1	2	1
1	1	2	1	2

$(3 \times 1) + (1 \times 2) = 5$ *This is not the number we are looking for.*

This is the correct combination.

3	2	1	2	1
1	1	2	1	2

$(3 \times 2) + (1 \times 1) = 7$ *This is the number we are looking for.*

The numbers on the top line of the table of combinations are the numbers for one bracket, and the bottom line gives the numbers for the other bracket.

3	2	1	2	1
1	1	2	1	2

So $3x^2 + 7x + 2 = (3x + 1)(x + 2)$

Check

	x	2
$3x$	$3x^2$	$6x$
1	x	2

So $(3x + 1)(x + 2) = 3x^2 + 6x + x + 2$
$ = 3x^2 + 7x + 2$

Example 2   Factorise $2x^2 − 5x − 3$.

The table of combinations is

2	3	−1	−3	1
1	−1	3	1	−3

All these pairs of numbers multiply to give −3 – put both ways round.

We now 'cross-multiply' and add, hoping to get 5.

2	3	−1	−3	1
1	−1	3	1	−3

$(2 \times −1) + (1 \times 3) = 1$ *This is not the number we are looking for.*

2	3	−1	−3	1
1	−1	3	1	−3

$(2 \times 3) + (1 \times -1) = 5$ — This is not the number we are looking for.

2	3	−1	−3	1
1	−1	3	1	−3

$(2 \times 1) + (1 \times -3) = -1$ — This is not the number we are looking for.

2	3	−1	−3	1
1	−1	3	1	−3

$(2 \times -3) + (1 \times 1) = -5$ — This is the number we are looking for.

This is the correct combination.

So $2x^2 - 5x - 3 = (2x + 1)(x - 3)$

Sometimes the coefficient of x^2 has more than one set of factors. If this is so, a separate table must be drawn for each set.

Example 3 Factorise $6x^2 + 7x - 3$.

Here there are two sets of factors of 6: 6 and 1, and 3 and 2.

The table for 6 and 1 is

6	−3	1	3	−1
1	1	−3	−1	3

We 'cross-multiply' and add, hoping to get 7.

6	−3	1	3	−1
1	1	−3	−1	3

$(6 \times 1) + (1 \times -3) = 3$ No

6	−3	1	3	−1
1	1	−3	−1	3

$(6 \times -3) + (1 \times 1) = -17$ No

6	−3	1	3	−1
	1	−3	−1	3

$(6 \times -1) + (1 \times 3) = -3$ No

6	−3	1	3	−1
1	1	−3	−1	3

$(6 \times 3) + (1 \times -1) = 17$ No

Clearly the factors 6 and 1 do not give us the correct combination. So now we try 3 and 2. We find that this is the correct combination.

3	−3	1	3	−1
2	1	−3	−1	3

$(3 \times 3) + (2 \times -1) = 7$

So $6x^2 + 7x - 3 = (3x - 1)(2x + 3)$

C2 Now try these.

(a) $2x^2 + 7x + 3$
(b) $2x^2 - 7x - 4$
(c) $3x^2 + 5x + 2$
(d) $5x^2 - 12x + 4$
(e) $3x^2 + 7x - 6$
(f) $8x^2 - 14x + 3$
(g) $8a^2 - 10a - 3$
(h) $6x^2 + 19x + 10$
(i) $4x^2 + 13x - 12$
(j) $4b^2 - 16b + 15$
(k) $4c^2 - 13c + 9$
(l) $10d^2 - d - 2$
(m) $6e^2 + 17e - 14$
(n) $8x^2 - 2x - 3$
(o) $10x^2 - 39x + 14$
(p) $3x^2 + 11x + 10$
(q) $4g^2 - g - 5$
(r) $6a^2 - 23a - 4$
(s) $2p^2 + 3p - 5$
(t) $12x^2 + 20x + 3$
(u) $3t^2 + t - 4$
(v) $6x^2 - 11x + 4$
(w) $10d^2 + 19d + 6$
(x) $24x^2 - 14x - 5$

D Other quadratic factorisation

Expressions like $-3x^2 + 10x - 8$, where the coefficient of the x^2 term is negative, can be factorised using the same method.

The table for $-3x^2 + 10x - 8$ is

Only use one set of factors here – we do not need 3 and −1 as well.

−3	8	−1	−8	1	4	−2	−4	2
1	−1	8	1	−8	−2	4	2	−4

The correct combination using 'cross-multiplication' and adding to get 10 is

−3	8	−1	−8	1	4	−2	−4	2
1	−1	8	1	−8	−2	4	2	−4

$(-3 \times -2) + (1 \times 4) = 10$

So $-3x^2 + 10x - 8 = (-3x + 4)(x - 2)$

Note that $(3x - 4)(-x + 2) = -3x^2 + 10x - 8$ also. This is the factorised form we obtain if we use 3 and −1 as the factors of −3.

D1 Check that $(3x - 4)(-x + 2) = -3x^2 + 10x - 8$.

Question D1 shows that another factorised form can be obtained by changing the sign of every term in the brackets. This is always true. For example,

$(5x + 1)(x - 9) = (-5x - 1)(-x + 9)$ and so on

D2 Factorise these.
(a) $-2x^2 - x + 3$
(b) $-2x^2 + 9x - 4$
(c) $-6x^2 + 11x - 4$
(d) $-3x^2 + 17x - 10$
(e) $-4x^2 + 11x - 6$
(f) $-6x^2 - 11x - 5$

Expressions like $3(x + 2)^2 + 7(x + 2) + 2$ can also be factorised using this method. Replace $(x + 2)$ by B.

Then $3(x + 2)^2 + 7(x + 2) + 2$ becomes $3B^2 + 7B + 2$.

The table is like this.

3	2	1	−2	−1
1	1	2	−1	−2

The correct combination using 'cross-multiplication' and adding is

3	2	1	−2	−1
1	1	2	−1	−2

$(3 \times 2) + (1 \times 1) = 7$

So $3B^2 + 7B + 2 = (3B + 1)(B + 2)$

Replace B by $(x + 2)$

$3(x + 2)^2 + 7(x + 2) + 2 = (3(x + 2) + 1)((x + 2) + 2)$
$= (3x + 6 + 1)(x + 2 + 2)$
$= (3x + 7)(x + 4)$

This technique can also be used with expressions like $2x^4 - x^2 - 6$.
Write $2x^4 - x^2 - 6$ as $2(x^2)^2 - x^2 - 6$. Replace x^2 by B. Then

$2x^4 - x^2 - 6$ becomes $2B^2 - B - 6$

which factorises as $(2B + 3)(B - 2)$.

Now replace B by x^2. So

$2x^4 - x^2 - 6 = (2x^2 + 3)(x^2 - 2)$

D3 Factorise these.

(a) $2(x + 1)^2 - 7(x + 1) - 4$
(b) $5(x + 7)^2 - 18(x + 7) - 8$
(c) $7(2x - 3)^2 - 10(2x - 3) + 3$
(d) $8(4 - 2x)^2 + 2(4 - 2x) - 6$
(e) $3x^4 - x^2 - 10$
(f) $4x^4 + 7x^2 - 2$
(g) $6x^4 - 5x^2 + 1$
(h) $12x^4 + 22x^2 + 8$
(i) $6(x - 2)^2 + 17(x - 2) - 3$
(j) $8x^6 - 10x^3 + 3$
(k) $2x^6 + 7x^3 - 4$
(l) $9(5 - x)^2 - 9(5 - x) + 2$

D4 Factorise these.

(a) $\sin^2 x + 4 \sin x + 3$
(b) $\cos^2 x + 5 \cos x + 6$
(c) $\sin^2 x + 3 \sin x - 10$
(d) $2 \sin^2 x - 5 \sin x + 2$
(e) $6 \sin^2 x + \sin x - 2$
(f) $4 \cos^2 x - 7 \cos x + 3$
(g) $6 \cos^2 x + 19 \cos x + 8$
(h) $8 \cos^2 x + 14 \cos x - 15$

D5 Factorise these.

(a) $x^2 - 2xy - 3y^2$
(b) $x^2 - 5xy + 6y^2$
(c) $2x^2 - 7xy - 4y^2$
(d) $6a^2 - ab - 2b^2$
(e) $4p^2 - 3pq - 10q^2$
(f) $3x^2 + 10xy + 3y^2$
(g) $6k^2 - 7kl + 2l^2$
(h) $8y^2 + 6yz - 9z^2$
(i) $6a^2 + 5ab - 4b^2$

3 Trigonometric graphs

A The big wheel

Richard and Alison have built a miniature big wheel for the model village of Weefolksberg.

The radius of the wheel is 1 m. Round the edge of the wheel are small cabins in which model people sit.

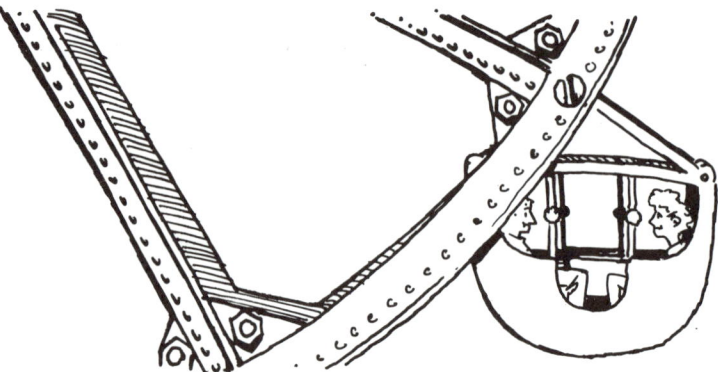

We are going to look at the path followed by one of these cabins as the wheel turns.

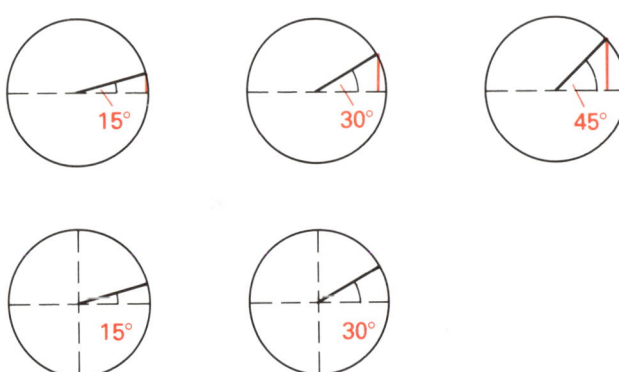

A1 (a) Copy and complete this table to show how the height of the cabin above or below the horizontal line through the centre changes as the wheel turns.

Angle of turning	0°	15°	30°	45°	60°	330°	345°	360°
Above or below	0 m	Above	Above					
Distance	0 m	0.26 m						

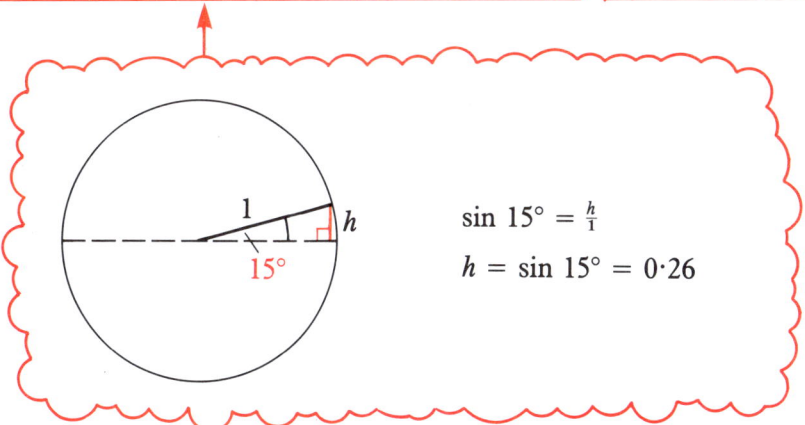

$\sin 15° = \frac{h}{1}$

$h = \sin 15° = 0.26$

(b) Copy and complete this graph from your table.

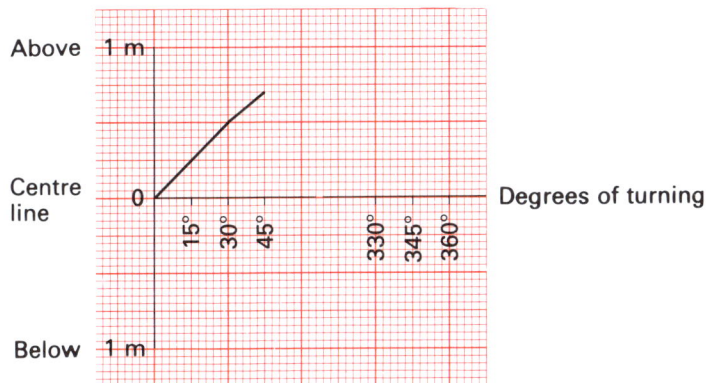

(c) What is the maximum height reached?
What is the minimum height reached?

(d) What would happen to the graph if the wheel kept turning?

(e) Can you find a rule which would allow you to work out the height above or below the horizontal for any angle of turning?

25

A2 (a) Copy and complete this table which shows how far the cabin is to the left or right of the vertical line through the centre.

Angle of turning	0°	15°	30°		330°	345°	360°
Right or left	Right						
Distance	1 m						

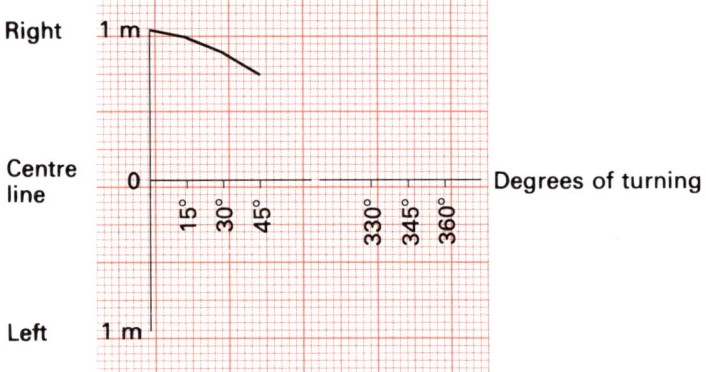

(b) Copy and complete this graph from your table.

(c) How far to the right does the cabin reach?
How far to the left does the cabin reach?

(d) What would happen to the graph if the wheel kept turning?

(e) Can you find a rule which would allow you to work out how far to the left or right the cabin is after any degree of turning?

A3 A full sized big wheel has radius 15 m.

(a) Sketch graphs to show, as the wheel turns, the distance of a cabin

(i) above or below the horizontal line through the centre

(ii) to the left or right of the vertical line through the centre

(b) What are the maximum and minimum heights reached by the cabin?

(c) What is the furthest right and left that the cabin reaches?

(d) Has the period of the graph changed?

A4 Think about what would happen for big wheels of differing radii.

Sketch some graphs of your results.

Would the maxima and minima change?

Would the periods change?

B **Change the range**

Here are the graphs we met first in the last section, together with their rules.

f(x) = sin x°   f(x) = cos x°

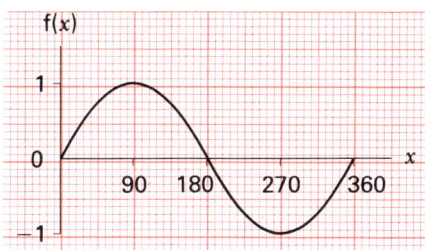 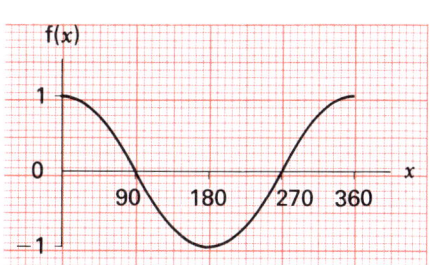

> f(x) = sin x° and f(x) = cos x°
> are periodic functions.
> The period is 360°.

Both graphs lie between 1 and −1. This is called the **range** of the function.

We shall now look at how we can change the range of the function, as we did in section A.

B1 Here is the graph of f(x) = 2 cos x° for 0 ≤ x ≤ 360.

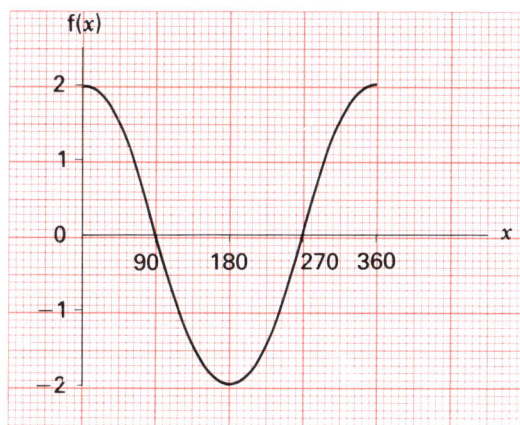

Compare this with the graph of f(x) = cos x°.

Write down for the following functions

    (i) the period   (ii) the maximum and minimum values

(a) f(x) = 4 cos x°   (b) f(x) = 10 sin x°   (c) f(x) = a cos x°.

27

B2 Here is the graph of $f(x) = \cos x° + 1$ for $0 \leq x \leq 360$.

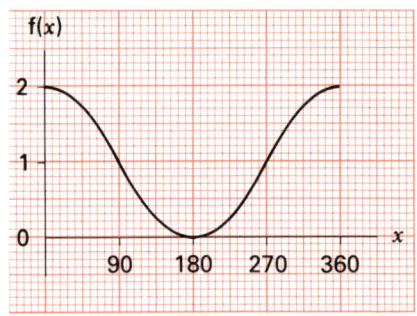

Compare this with the graph of $f(x) = \cos x°$.

Explain what has happened to (a) its period (b) the maximum and minimum values.

B3 Write down (i) the period (ii) the maximum and minimum values of the following functions.

(a) $f(x) = \cos x° + 4$ (b) $f(x) = \cos x° - 3$ (c) $f(x) = \sin x° + 5$.

C Change the period

C1 Here is the graph of the function $f(x) = \sin 2x°$ for $0 \leq x \leq 360$.

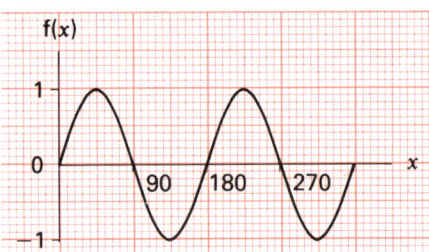

Compare this with the graph of $f(x) = \sin x°$.

(a) What has happened to the period of the graph?

(b) What has happened to its maximum and minimum values?

C2 Here is the graph of $f(x) = \sin 4x°$ for $0 \leq x \leq 360°$.

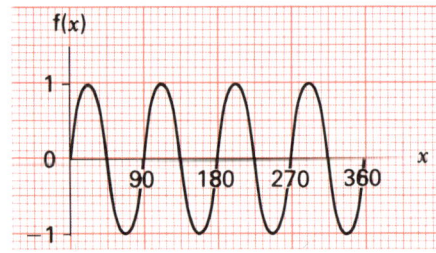

Compare this with the graph of $f(x) = \sin x°$.

(a) What has happened to the period of the graph?

(b) What has happened to its maximum and minimum values?

C3 Write down the periods of the following graphs, without drawing them.

(a) $f(x) = \sin 3x°$ (b) $f(x) = \sin 10x°$ (c) $f(x) = \sin ax°$

(d) What can you say about their maximum and minimum values?

C4 Sketch the following graphs.

(a) $f(x) = 5 \sin 2x°$ (b) $f(x) = 7 \cos 3x° + 4$

C5 Write down the period and the maximum and minimum value of the function.

$f(x) = a \cos bx° + 1$

D Change the roots

D1 Here is part of the graph of the function $f(x) = \sin x°$.

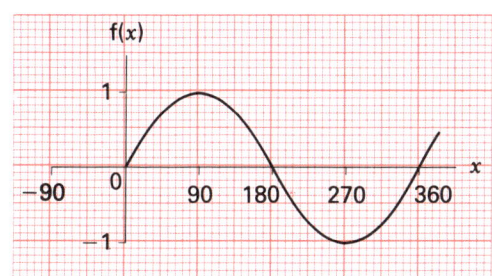

Here is part of the function $f(x) = \sin (x - 90)°$

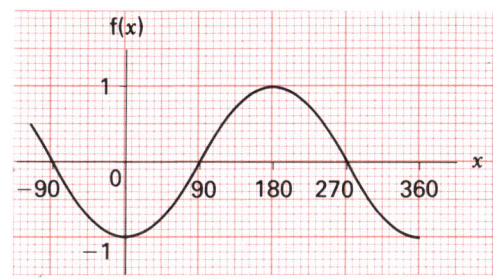

Explain clearly what has happened to the graph of sin $x°$.

D2 Sketch the graph of sin $(x - 180)°$.

D3 Sketch the graph of sin $(x + 90)°$.

D4 Write down another rule for a graph in the same position as $f(x) = \sin x°$.

D5 For each of the following graphs, give

      (i) its period   (ii) its maximum and minimum values

  (a) $f(x) = 3 \cos 4x°$ (b) $f(x) = \sin (x - 45)°$

  (c) $f(x) = 4 \cos (x + 90)° - 3$

D6 For $0 \leq x \leq 360$, write down the roots of the graphs in question D5.

D7 Sketch the following graphs for $0 \leq x \leq 360$

  (a) $f(x) = 2 \sin 3x°$ (b) $f(x) = 5 \cos x° + 2$

  (c) $f(x) = 2 \sin (x - 90)°$

D8 Write down suitable rules for these graphs.

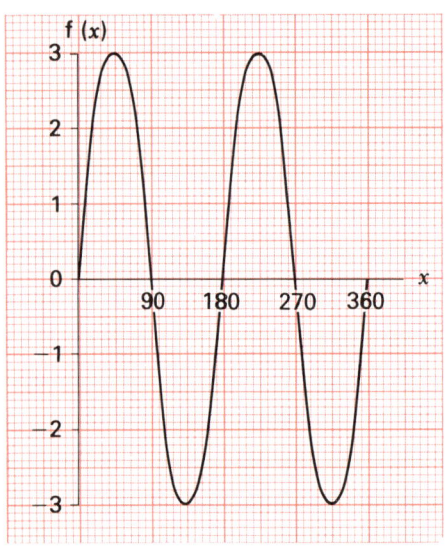
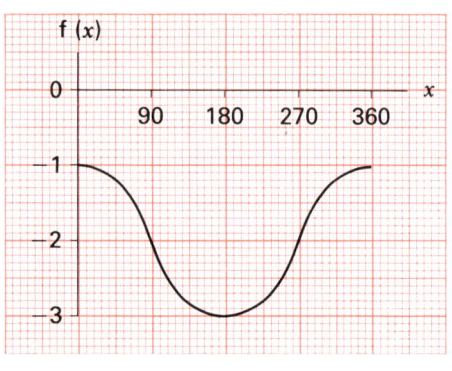

E An all-encompassing problem

Investigate how the distance d cm between the ends of a pair of compasses changes as the angle $a°$ changes.

Illustrate your answers with a graph.

Can you write down a rule for the graph you have drawn?

Investigate what happens to the rule for different sizes of compasses.

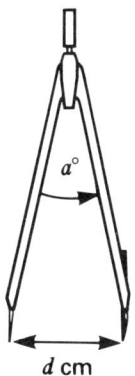

A Difference of two squares

A1 Explain how you could work out $49^2 - 39^2$ without showing any working (and without using a calculator!).

A2 Factorise the following.

(a) $25b^2 - 36$
(b) $4x^2y^2 - 9y^2z^2$

(c) $81x^4y^2 - 49z^2$
(d) $200a^4b^6 - 72b^2c^4$

(e) $(4 + a)^2 - b^2$
(f) $(3x + 2)^2 - (x - 9)^2$

(g) $(a + b + c)^2 - (a + b - c)^2$

A3 From this rectangular metal plate, a circular hole of radius r cm is cut.

(a) Write an expression for the area of the shaded region.

(b) Write this expression in fully factorised form.

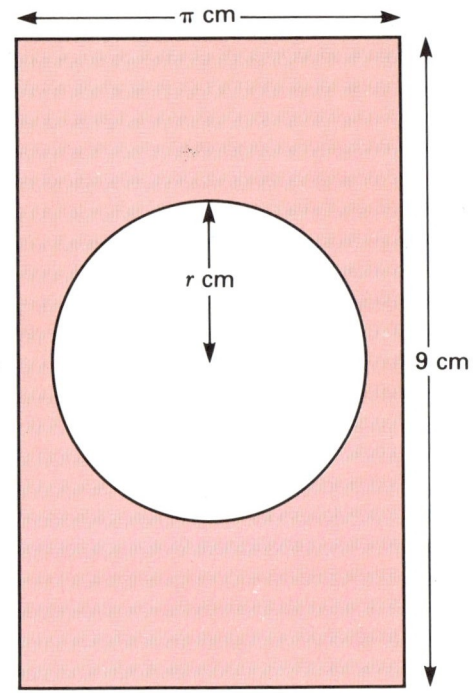

A4 A glazier has two square sheets of glass, one of which has side 5 cm longer than the other.

(a) If the length of the smaller sheet is x cm, write an expression for the difference in their areas.

(b) Factorise your answer.

B Mortgages

B1 Robert wants to buy this house priced at £38 000.

His building society will lend him up to 2·85 times his annual salary or 95% of the value of the house, whichever is less. Robert's salary is £12 530.

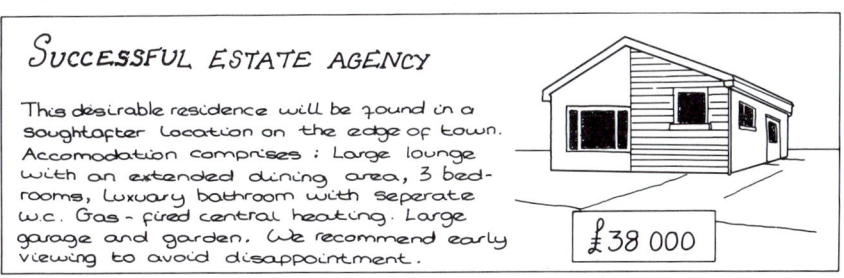

> SUCCESSFUL ESTATE AGENCY
>
> This desirable residence will be found in a soughtafter location on the edge of town. Accomodation comprises : Large lounge with an extended dining area, 3 bedrooms, luxury bathroom with seperate w.c. Gas-fired central heating. Large garage and garden. We recommend early viewing to avoid disappointment.
>
> £38 000

(a) What would be the maximum amount that the society would lend him, based on the value of the property?

(b) What would be the maximum that they would lend him based on his salary?

(c) What would be the maximum mortgage the society would lend Robert on the property?

B2 Mary goes to the same building society as Robert. She has been granted a mortgage based on her salary as long as she pays a deposit of £2575. The house is valued at £32 500.

What is Mary's annual salary?

B3 Alan and Susan can pay their mortgage in monthly installments of £283·55 for 20 years or £245·59 for 25 years.

Which method will result in them paying the least and by how much?

C Factors of quadratics

C1 Factorise the following.

(a) $a^2 + 3a + 2$

(b) $2b^2 + 7b + 3$

(c) $6c^2 - 5c - 6$

(d) $16d^2 + 14d - 30$

(e) $14e^2 + e - 4$

(f) $-6a^2 - a + 2$

(g) $x^2 + 3xy + 2y^2$

(h) $12x^2 - 17xy - 5y^2$

(i) $8a^2 + 2ab - 3b^2$

(j) $10(x + 3)^2 - 7(x + 3) - 3$

(k) $8a^4 + 14a^2 - 15$

(l) $12x^4 + 21x^2 - 6$

(m) $12 \sin^2 x + 5 \sin x - 2$

(n) $-6(x + 2)^2 - 5(x + 2) + 4$

(o) $-4x^4 + 16x^2 - 15$

C2 A rectangular piece of wood has an area of $(x^2 - 4x - 45)$ cm². If its length is $(x + 5)$ cm, find its width.

C3 This box has a square base, and height 3 cm.
Its volume is $(3x^2 + 24x + 48)$ cm³.
Find an expression for the length of the base in terms of x.

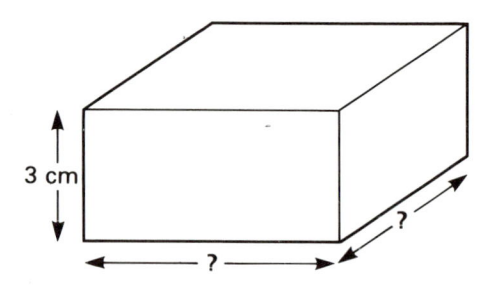

C4 If $x - 1$ is a factor of $x^2 + 3x + c$, find c.

C5 If $x + 3$ is a factor of $6x^2 + bx - 15$, find b.

D Trigonometric graphs

D1 Sketch the graphs of the following functions for $0 \leqslant x \leqslant 360$. Show clearly the period and maximum and minimum values.

(a) $f(x) = 5 \sin 2x°$
(b) $f(x) = 2 \cos 3x°$
(c) $f(x) = 2 \sin 2x°$

D2 Here is a pair of dividers. The distance between the ends of the legs changes as the dividers are opened up.

(a) Copy and complete the following table for the dividers.

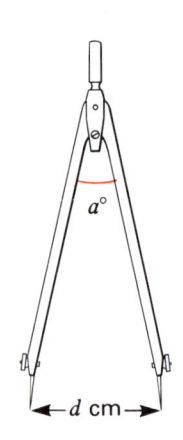

a	0	20	40	60	...	180
d	0	2·78				

(b) Sketch the graph of (a, d) for $0 \leqslant a \leqslant 180$.

(c) Write down a rule for calculating d for any angle $a°$.
Is there any restriction on the size of $a°$?

4 Indices

You already know that expressions like $2 \times 2 \times 2 \times 2$ and $5 \times 5 \times 5 \times 5 \times 5$ can be written in short as 2^4 and 5^5.

The same can be done with letters:

$a \times a \times a \times a = a^4$
$b \times b \times b = b^3$
$c \times c \times c \times c \times c = c^5$

These numbers are called **powers** or **indices**.

This chapter deals with laws of indices which we shall try to discover.

A Products of powers

Examples

$3^3 = 3 \times 3 \times 3$
$3^2 = 3 \times 3$
$3^3 \times 3^2 = (3 \times 3 \times 3) \times (3 \times 3)$
$ = 3^5$

$a^3 = a \times a \times a$
$a^2 = a \times a$
$a^3 \times a^2 = (a \times a \times a) \times (a \times a)$
$ = a^5$

A1 In the same way, work out a short way of writing

(a) $5^2 \times 5^4$ (b) $6^3 \times 6^4$ (c) $8^5 \times 8^6$ (d) $b^2 \times b^5$

(e) $c^7 \times c^3$ (f) $d^4 \times d^9$ (g) $e^7 \times e^6 \times e^2$

A2 Use your answers to question A1 to explain what happens when we multiply powers.

A3 Use this result to write down an expression for $a^x \times a^y$.

This result works whenever we have powers of the same number. It is called a **product rule**.

A4 Work out the following. The first one has been done for you.

(a) $a^7 \times b^3 \times a^9 \times b^2 = a^{16} b^5$

(b) $c^2 \times c^3 \times d^4 \times c^5 \times d^6$

(c) $e^4 \times f^2 \times g^3 \times g^2 \times f^9$

(d) $x^9 \times y^3 \times x^2 \times z^6 \times y^7 \times z^3$

Examples

$3^2 \times 2^2 = 9 \times 4 = 36 = 6^2$

$(3 \times 2)^2 = 6^2 = 36$

Equating these, we have $3^2 \times 2^2 = (3 \times 2)^2$

$(ab)^2 = (a \times b)^2$
$= (a \times b) \times (a \times b)$
$= (a \times a) \times (b \times b)$
$= a^2 \times b^2$
$= a^2b^2$

A5 Explain this rule in words.
In general, we can write $a^x b^x = (ab)^x$
This is another product rule.

A6 Use the above rule to simplify the following. The first one has been done for you.

(a) $4^3 \times 5^3 = (4 \times 5)^3 = 20^3 = 8000$

(b) $5^2 \times 10^2$ (c) $2^4 \times 5^4$ (d) $3^3 \times 5^3$ (e) $a^4 b^4$

(f) $3^3 b^3$ (g) $p^6 q^6 r^6$

B Dividing powers

$\dfrac{5^5}{5^3} = \dfrac{5 \times 5 \times 5 \times 5 \times 5}{5 \times 5 \times 5}$ *These cancel out.*

$= 5 \times 5$
$= 5^2$

$\dfrac{a^5}{a^3} = \dfrac{a \times a \times a \times a \times a}{a \times a \times a}$

$= a \times a$
$= a^2$

B1 In the same way, simplify the following.

(a) $\dfrac{9^7}{9^3}$ (b) $\dfrac{4^8}{4^6}$ (c) $\dfrac{12^5}{12^4}$ (d) $\dfrac{e^9}{e^3}$ (e) $\dfrac{f^2}{f^5}$ (f) $\dfrac{g^6}{g^5}$

B2 Explain in words what happens when we divide powers.

B3 Use this result to write down an expression for $\dfrac{a^x}{a^y}$.

This is called the **quotient rule** for indices.

B4 Use this result to simplify the following.

36

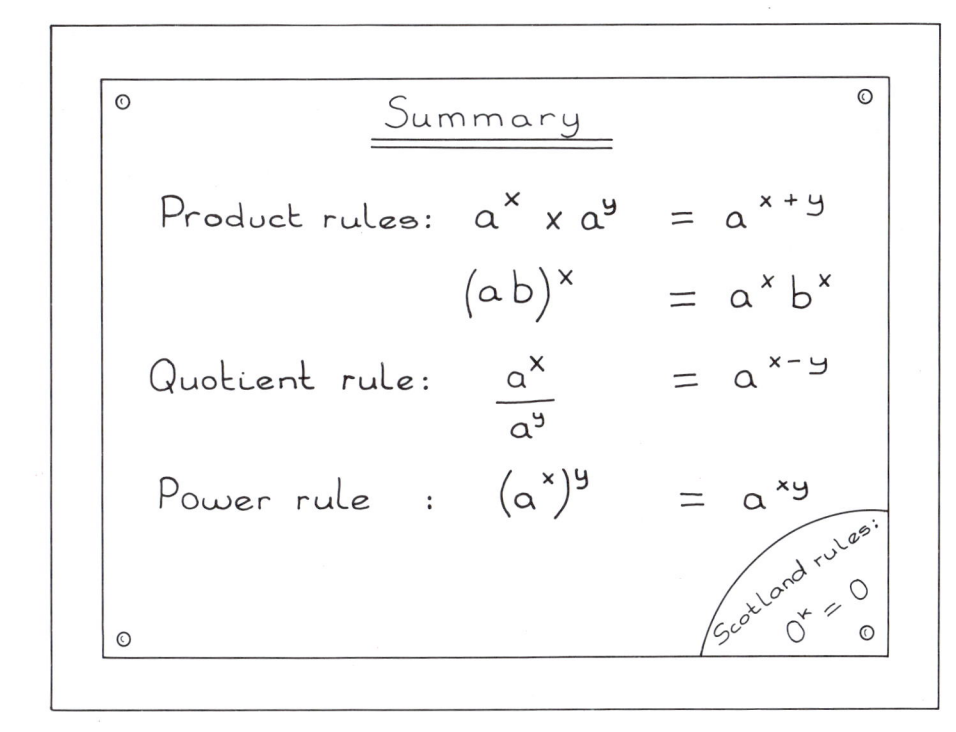

(a) $\dfrac{x^7}{x^5}$ (b) $\dfrac{y^6}{y^2}$ (c) $\dfrac{z^3}{z^2}$ (d) $\dfrac{x^7 y^5}{x^3 y^2}$ (e) $\dfrac{9a^{13}}{27a^2}$

C Powers of powers

C1 Copy and complete the following.

(a) $(7^2)^2 = (7 \times 7) \times (7 \times 7) = 7^?$

(b) $(a^2)^3 = (a \times a) \times (a \times a) \times (a \times a) = a^?$

C2 In the same way, simplify the following

(a) $(5^3)^5$ (b) $(b^3)^2$ (c) $(c^2)^4$ (d) $(d^5)^2$

C3 Explain what happens in words

C4 Use this result to write down an expression for, $(a^x)^y$.
This is called the **power rule** for indices.

C5 Use your result to simplify the following.

(a) $(x^4)^3$ (b) $(y^5)^3$ (c) $(z^9)^2$ (d) $(a^7)^4$
(e) $(3a^2)^3$ (f) $(2b^3)^4$ (g) $(5c^4)^2$ (h) $(7d^9)^3$

Summary

Product rules: $a^x \times a^y = a^{x+y}$

$(ab)^x = a^x b^x$

Quotient rule: $\dfrac{a^x}{a^y} = a^{x-y}$

Power rule : $(a^x)^y = a^{xy}$

Scotland rules: $0^k = 0$

D Indices less than or equal to zero

This table shows the values of differing powers of 2.

n	1	2	3	4	5
2^n	2	4	8	16	32

×2 ×2 ×2 ×2

Extending the table backwards we have

n	5	4	3	2	1	0	−1	−2	−3
2^n	32	16	8	4	2	1	$\frac{1}{2}$	$\frac{1}{4}$	$\frac{1}{8}$

From this table we can see that

$2^0 = 1$
$2^{-1} = \frac{1}{2} = \frac{1}{2^1}$
$2^{-2} = \frac{1}{4} = \frac{1}{2^2}$
$2^{-3} = \frac{1}{8} = \frac{1}{2^3}$

D1 Copy and complete this table for powers of 3.

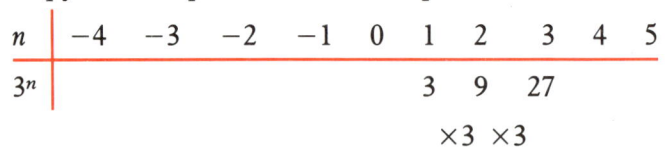

n	−4	−3	−2	−1	0	1	2	3	4	5
3^n							3	9	27	

×3 ×3

Leave your answers as fractions where necessary.

D2 Copy and complete this table for powers of 5.

n	−3	−2	−1	0	1	2	3
5^n							

Leave your answers as fractions where necessary.

D3 What can you say about 3^0 and 5^0?

D4 Try to find a relationship between 2^{-3} and 2^3, 3^{-2} and 3^2, 5^{-1} and 5^1.

D5 (a) Write down a general rule for a^0.

(b) Write down a general rule for a^{-n} and a^n.

D6 Write down the value of

(a) 7^0 (b) 32^0 (c) 7563^0 (d) x^0 (e) $(y^4)^0$

D7 Write each of the following using negative indices.

(a) $\dfrac{1}{9^4}$ (b) $\dfrac{1}{5^7}$ (c) $\dfrac{1}{c^5}$ (d) $\dfrac{1}{a^{11}}$ (e) $\dfrac{2}{e^2}$ (f) $\dfrac{5}{g^{12}}$ (g) $\dfrac{1}{a^x}$

D8 Write each of the following as fractions.

(a) 7^{-3} (b) 12^{-6} (c) a^{-3} (d) y^{-8} (e) $(3x)^{-2}$ (f) $\tfrac{1}{2}x^{-7}$ (g) a^{-x}

E Fractional indices

We shall now look at how we can write expressions like $\sqrt{2}, \sqrt[3]{7}$ using indices.

We already know that $(\sqrt{2})^2 = 2 = 2^1$

Using the power rule for indices, $2^1 = 2^{\frac{1}{2} \times 2} = (2^{\frac{1}{2}})^2$

Equating these we have
$\sqrt{2} = 2^{\frac{1}{2}}$

Here is another example.

$(\sqrt[3]{a})^3 = a = a^1$
$a^1 = a^{\frac{1}{3} \times 3} = (a^{\frac{1}{3}})^3$

Equating these we have $\sqrt[3]{a} = a^{\frac{1}{3}}$

As a general rule we can write

$a^{\frac{1}{n}} = \sqrt[n]{a}$

We say 'the nth root of a'.

E1 Write the following using root signs.

The first one has been done for you.

(a) $x^{\frac{1}{4}} = \sqrt[4]{x}$ (b) $y^{\frac{1}{6}}$ (c) $a^{\frac{1}{3}}$ (d) $c^{\frac{1}{5}}$

E2 Evaluate the following, rounding your answers to two decimal places where necessary.

(a) $25^{\frac{1}{2}}$ (b) $8^{\frac{1}{3}}$ (c) $3^{\frac{1}{2}}$ (d) $216^{\frac{1}{3}}$ (e) $11^{\frac{1}{2}}$ (f) $92^{\frac{1}{2}}$

E3 Express the following using fractional indices.

(a) $\sqrt[3]{b}$ (b) $\sqrt[7]{x}$ (c) $\sqrt[5]{a}$ (d) $\sqrt{9a}$

Examples

$a^{\frac{3}{4}} = (a^3)^{\frac{1}{4}} = \sqrt[4]{a^3}$

$x^{\frac{2}{5}} = (x^2)^{\frac{1}{5}} = \sqrt[5]{x^2}$

Using the power rule of indices

In general we can write

$a^{\frac{x}{y}} = \sqrt[y]{a^x}$

E4 Write the following in root form.

(a) $a^{\frac{3}{5}} = \sqrt[5]{a^3}$ (b) $x^{\frac{1}{4}}$ (c) $d^{\frac{2}{5}}$ (d) $p^{\frac{9}{4}}$

The first one has been done for you.

E5 Express the following using fractional indices.

(a) $\sqrt{b^3}$ (b) $\sqrt[5]{q^2}$ (c) $\sqrt[3]{x^5}$ (d) $\sqrt[5]{a^9}$ (e) $\sqrt[7]{a^5}$

E6 Simplify the following using the product, quotient and power rules of indices.

(a) $a^{\frac{3}{4}} \times a^{\frac{1}{4}}$ (b) $(a^{-3})^9$ (c) $\dfrac{a^{-6}}{a^3}$ (d) $a^{-\frac{1}{2}}$

(e) $\dfrac{a^{-3}}{a^{-3}}$ (f) $(a^{-9})^{\frac{1}{3}}$ (g) $(a^{-\frac{1}{4}})^{\frac{1}{3}}$ (h) $\dfrac{16a^{\frac{1}{2}}}{a^{-3}}$

F Miscellany

You have now met the following rules of indices.

$a^x \times a^y = a^{x+y}$ $(ab)^x = a^x b^x$

$\dfrac{a^x}{a^y} = a^{x-y}$ $a^{-x} = \dfrac{1}{a^x}$

$(a^x)^y = a^{xy}$ $a^{\frac{x}{y}} = \sqrt[y]{a^x}$

Use these rules to simplify the following.

F1 $(a^7)^3$ **F2** $(ab^3)^4$ **F3** $x^9(x^3)^2$ **F4** $\dfrac{(x^2y^3)^4}{(xy^2)^2}$

F5 $(\sqrt{a})^3$ **F6** $(x\sqrt{y})^4$ **F7** $\dfrac{1}{x^{-2}}$ **F8** $\dfrac{x^3}{x^{-3}}$

F9 $\dfrac{(6y)^2}{(2y)^3}$ **F10** $\dfrac{\sqrt[3]{b^2}}{b^3}$ **F11** $\dfrac{b^3}{\sqrt[3]{b^2}}$ **F12** $\dfrac{(-2yz^3)^2}{(-2z\sqrt{y})^3}$

G Equations with indices

Examples

$2^x = 16$
$2^x = 2^4$
$x = 4$

$9^x = 27$
$(3^2)^x = 3^3$
$3^{2x} = 3^3$
$2x = 3$
$x = \frac{3}{2}$

Solve the following equations.

G1 $5^x = 125$ **G2** $3^x = 81$ **G3** $4^x = 16$

G4 $16^x = 64$ **G5** $8^x = 256$ **G6** $1000^x = 1\,000\,000$

H Standard form

This is a common use of indices and is a method of writing very large numbers (like 125 000 000 000) and very small numbers (like 0·000 000 000 003) more concisely. We write the numbers in the form

> A number between 1 and 10, which can be 1, but not 10

× 10

> A positive or negative whole number.

Examples

The distance of Saturn from the Sun is 1 427 000 000 000 m. This is

$1·427 \times 1\,000\,000\,000\,000 = 1·427 \times 10^{12}$ m

The diameter of a white blood cell is 0·0004 cm, which is

$\dfrac{4}{10\,000} = \dfrac{4}{10^4} = 4 \times 10^{-4}$ cm

H1 Copy and complete this table.

Object	Size	Standard form
Radius of Earth	6 378 000 m	
Length of Venus orbit		$1·1 \times 10^9$ km
Diameter of polio virus		2×10^{-5} cm
Radius of liver cell	0·001 mm	

Rewrite the following numbers in standard form (rounded to 3 s.f. where necessary).

H2 32 507 000 **H3** 0·000 34 **H4** 12 456

H5 0·000 201 **H6** 8 701 000 **H7** 0·000 000 041 16

Numbers in standard form can be combined using the product and quotient rules of indices, like this:

$(3 \cdot 2 \times 10^4) \times (9 \cdot 1 \times 10^3)$
$= (3 \cdot 2 \times 9 \cdot 1) \times (10^4 \times 10^3)$ — Rearrange the order
$= 29 \cdot 12 \times 10^7$
$= 2 \cdot 912 \times 10^1 \times 10^7$ — Product rule of indices
$= 2 \cdot 912 \times 10^8$

$\dfrac{(4 \cdot 5 \times 10^3)}{(9 \cdot 0 \times 10^4)} = 0 \cdot 5 \times 10^{-1} = 5 \times 10^{-1} \times 10^{-1} = 5 \times 10^{-2}$ — Quotient rule of indices; Product rule of indices

H8 Copy and complete:

$(4 \cdot 5 \times 10^7) \times (3 \cdot 5 \times 10^4)$
$= (4 \cdot 5 \times \ldots) \times (10^7 \times \ldots)$
$= \ldots$

H9 $(6 \cdot 9 \times 10^3) \times (2 \cdot 8 \times 10^5)$ **H10** $(2 \cdot 8 \times 10^4) \times (6 \cdot 2 \times 10^{-3})$

H11 $(7 \cdot 8 \times 10^{-3}) \times (4 \cdot 8 \times 10^{-6})$ **H12** $(6 \cdot 32 \times 10^6) \times (5 \cdot 67 \times 10^{-5})$

H13 $\dfrac{(9 \cdot 6 \times 10^8)}{(3 \times 10^6)}$ **H14** $\dfrac{(2 \cdot 8 \times 10^5)}{(1 \cdot 4 \times 10^{-3})}$ **H15** $\dfrac{(1 \cdot 5 \times 10^2)}{(4 \cdot 5 \times 10^{-3})}$

I Problems

In these examples you have to decide whether to multiply or divide before you start.

Examples

The speed of light in space is $3 \cdot 00 \times 10^8$ metres per second. Rewrite this in km per hour.

$3 \cdot 00 \times 10^8$ m/s
$= 3 \cdot 00 \times 10^8 \times 60 \times 60$ m/h
$= 108 \times 10^{10}$ m/h
$= 108 \times 10^7$ km/h
$= 1 \cdot 08 \times 10^2 \times 10^7$ km/h
$= 1 \cdot 08 \times 10^9$ km/h

I1 The half-life of polonium is 3×10^{-7} seconds. Give this in the form $a \times 10^n$ years.

I2 The speed of light in space is $3\cdot00 \times 10^8$ metres per second.

How long does light take to reach Earth from the Sun ($1\cdot5 \times 10^{11}$ m away)?

I3 Cork weighs $0\cdot24 \times 10^3$ kg per cubic metre.

A wine-bottle cork has a volume of 16 cm³.
Work out the weight in grams of a wine-bottle cork.

I4 Bacteria have a radius of $3\cdot1 \times 10^{-6}$ cm. Calculate the total length if 12×10^6 of them could be placed side by side.

I5 The volume of a sphere is given by the formula

$$V = \tfrac{4}{3}\pi r^3$$

where r is the radius of the sphere.

(a) Calculate the volume of the Earth if its radius is $6\cdot4 \times 10^3$ km. Give your answer correct to 3 s.f.

(b) The mass of the Earth is calculated using the rule

mass = volume × density

Calculate the mass of the Earth if its density is $5\cdot5 \times 10^3$ kg per cubic metre.

5 Graphs 2

A Quotients

A1 Margaret is driving a distance of 200 km.

(a) Write down a rule you can use to work out how long the journey takes when you know her average speed.

(b) Use your rule to complete this table.

Speed (km/h)	20	40	60	80	100	120	140	150
Time (h)								

(c) Draw the graph of (time, speed).

A2 These rectangles all have the same area, 48 cm².

48 cm, b

8 cm, b

16 cm, b

These rectangles are **not** drawn to scale!

2 cm, b

6 cm, b

10 cm, b

(a) Write down a rule you can use to work out the breadth of a rectangle when you know its length.

(b) Use your rule to complete this table.

Length (cm)	2	6	8	10	16	48
Breadth (cm)						

(c) Draw the graph of (length, breadth). Add some other values to the table to help.

A3 (a) Make up a table of values for the function

$$f(x) = \frac{10}{x}$$

for values of x from -5 to 5.

(b) Why is there no value of f(0)?

(c) Draw the graph of f(x) for $-5 \leq x \leq 5$.

A4 Repeat question A3 for

$$f(x) = \frac{2}{x}$$

with $-4 \leq x \leq 4$.

A5 What do you notice about all the graphs in this section? Can you explain why some of them have only one branch?

B Exponential graphs

B1 These microscopic organisms reproduce by splitting into two every minute.

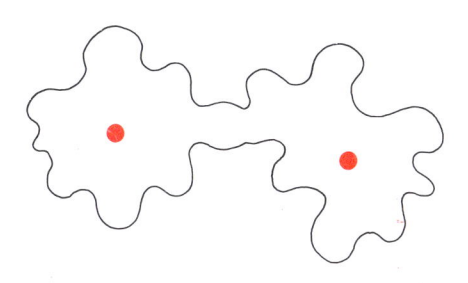

(a) Copy and complete this table to show how many organisms there are after each minute.

Time (min)	0	1	2	3	4	5	6
Organisms	1						

(b) Write down a rule for the number of organisms after n minutes.

(c) If you had started with two organisms, how many would there be after n minutes?

B2 Imagine an organism which reproduces by splitting into three every minute.

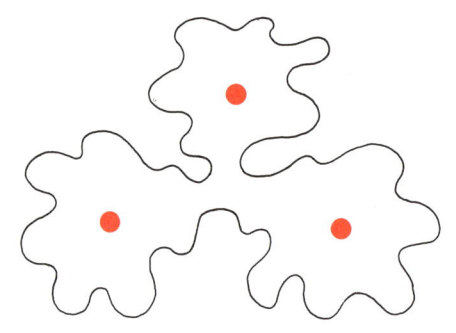

(a) Write down a rule for the number of organisms there would be after n minutes.

(b) If there had been five organisms to start with, how would your rule change?

Something which is multiplied by the same amount in equal periods of time is said to change **exponentially**.

For example, the number of organisms in question C2 increases exponentially. It is multiplied by 3 every time.

B3 Weed covers part of the surface of a pond. Every year the area covered doubles.

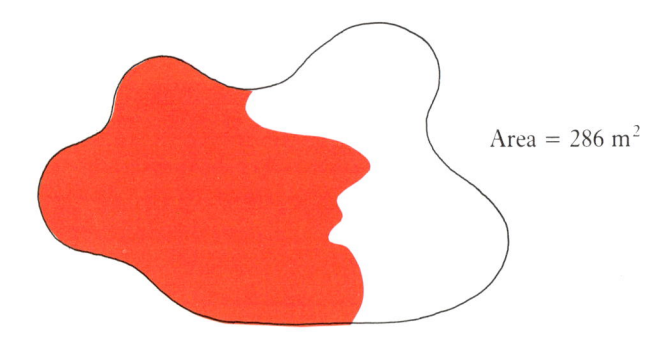

Area = 286 m^2

(a) Copy and complete this table.

Time (years)	0	1	2	3	4	5	6
Area (m²)	1	2					

(b) Draw the graph of (time, area)

(c) The pond has an area of 286 m². After how many years is more than half of the pond covered by weed?

(d) Find a rule for the amount of weed covering the pond in n years. What can you say about n?

B4 This car costs £6500 when new. Its value depreciates by 18% per year.

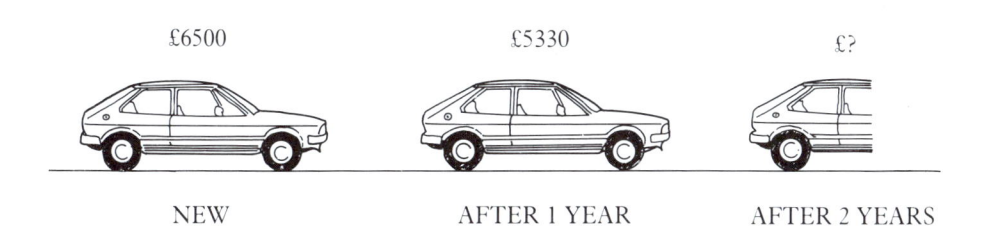

£6500   £5330   £?

NEW   AFTER 1 YEAR   AFTER 2 YEARS

(a) Copy and complete this table.

Age	0	1	2	3	4	5
Value	£6500	£5330				

× 0·82 × 0·82

The value of the car decreases exponentially. It is multiplied by 0·82 each time.

(b) Draw the graph of (age, value)

(c) After how many years does the car lose half its value?

(d) Find a rule for the value of the car after n years. (**Hint.** Look at the multipliers.)

B5 Carol has invested £100 in a bank account which pays compound interest of 10% annually.

Copy and complete this table.

Time (years)	0	1	2	3	4	5
Amount (£)	100	110				

　　　　　　　　　　x ?　x ?

(b) Draw the graph of (time, amount)

(c) How long would it take Carol to double her money?

(d) Try to find a rule for the amount of money Carol has in the account after n years.

Here are some questions about functions which increase or decrease exponentially.

B6 (a) Copy and complete the table for the function
$$f(x) = 2^x$$

x	0	1	2	3	4	5	6
$f(x)$	1	2	4				

(b) Draw the graph of $(x, f(x))$ for $0 \leqslant x \leqslant 6$.

B7 (a) Copy and complete the table for the function
$$f(x) = 3^x$$

x	−3	−2	−1	0	1	2	3
$f(x)$							

(b) Draw the graph of $(x, f(x))$ for $-3 \leqslant x \leqslant 3$.

B8 (a) Copy and complete the table for the function
$$f(x) = 5(2^x)$$

x	−2	−1	0	1	2	3	4
$f(x)$							

(b) Draw the graph of $(x, f(x))$ for $-2 \leqslant x \leqslant 4$.

(c) What are the values of $f(x)$ multiplied by each time x is increased by 1?

6 Similar areas and volumes

A Squares

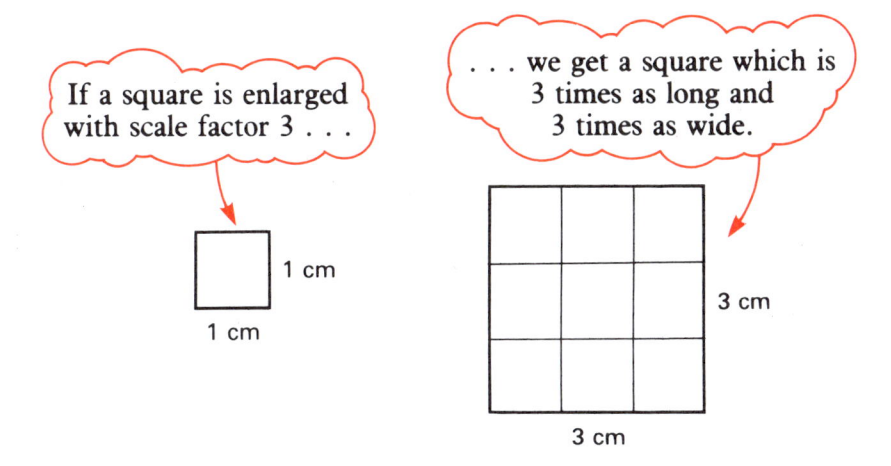

If a square is enlarged with scale factor 3 ...

... we get a square which is 3 times as long and 3 times as wide.

A1 (a) Write down the areas of the small and the enlarged square.

(b) By how many times has the area of the smaller square been enlarged?

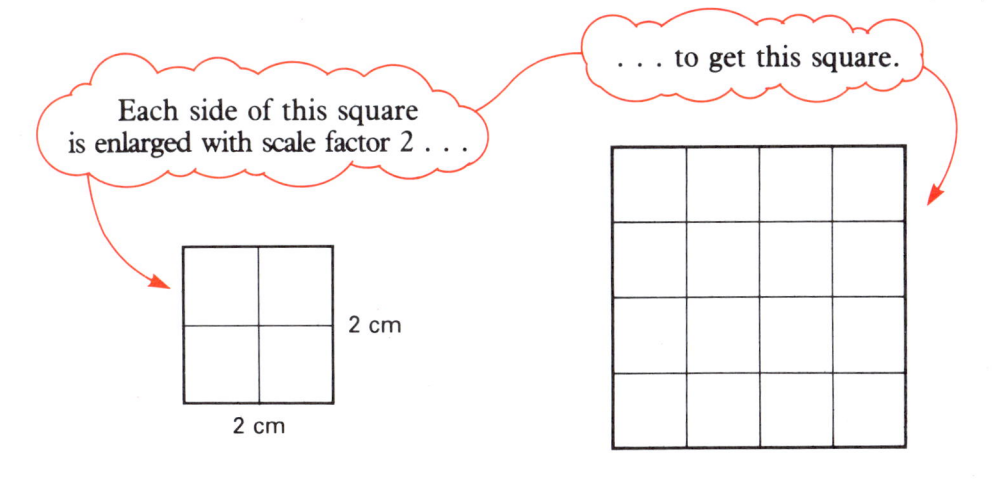

Each side of this square is enlarged with scale factor 2 ...

... to get this square.

This is called the **area factor**.

A2 (a) Write down the area of each square.

(b) Copy and complete:
The area of the larger square is __ times the area of the smaller square.

49

A3 (a) Write down the areas of these squares.

(b) Copy and complete:
The area of the smaller square is ___ times the area of the larger square.

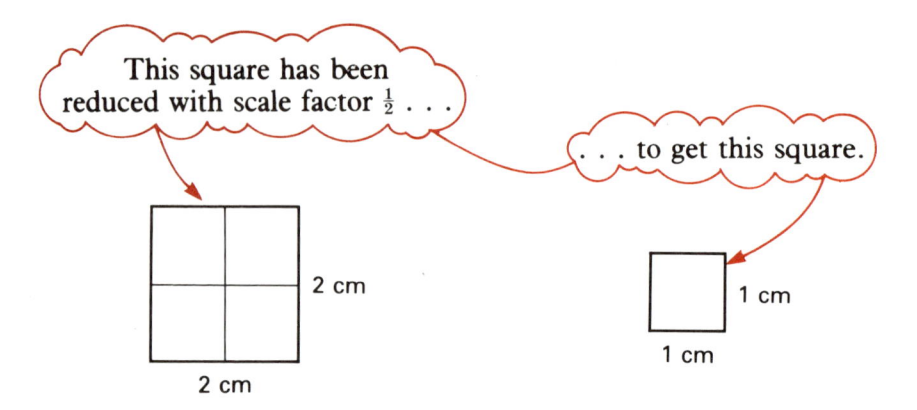

This square has been reduced with scale factor ½ ...

... to get this square.

2 cm

2 cm

1 cm

1 cm

A4 Write down the area factor for each of the following pairs of squares.

(a)

1 cm

1 cm

2 cm

2 cm

(b)

3 cm

3 cm

1 cm

1 cm

(c)

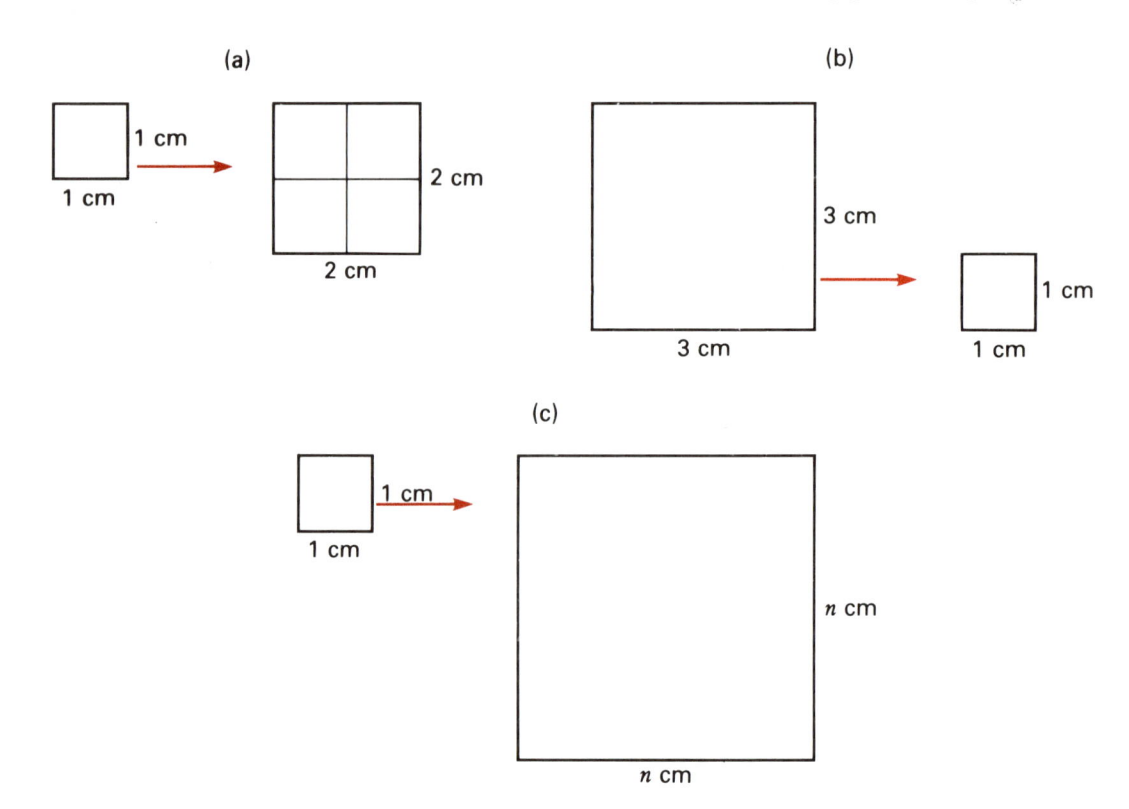

1 cm

1 cm

n cm

n cm

A5 Copy and complete this table.

Scale factor	Area factor
4	
	36
2·5	
	1·25

B Other areas

So far we have found that when enlarging or reducing squares:

- if the scale factor is 2, the area factor is 4, or 2^2,
- if the scale factor is 3, the area factor is 9, or 3^2
- if the scale factor is $\frac{1}{2}$, the area factor is $\frac{1}{4}$, or $(\frac{1}{2})^2$.

In general we can say

Area factor = (Scale factor)^2

This rule is true whatever area is enlarged or reduced.

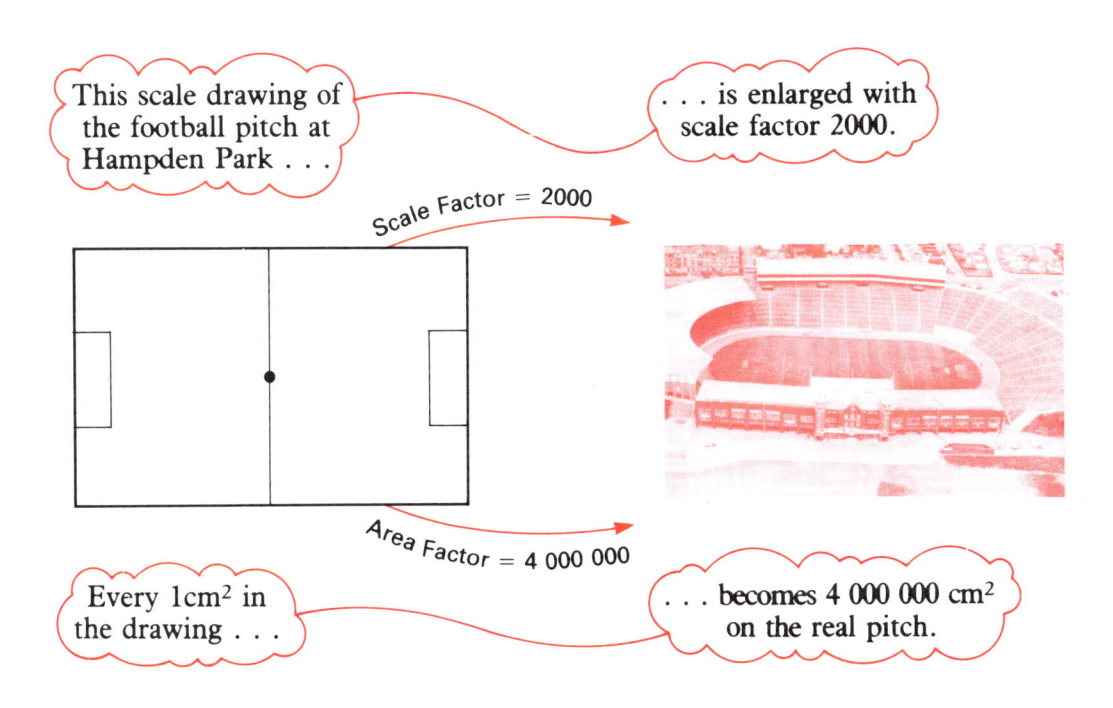

This scale drawing of the football pitch at Hampden Park . . .

. . . is enlarged with scale factor 2000.

Scale Factor = 2000

Area Factor = 4 000 000

Every 1cm^2 in the drawing . . .

. . . becomes 4 000 000 cm^2 on the real pitch.

B1 If the area of the scale drawing is 18·28 cm², calculate the area of the real pitch.

B2 A drawing is to be made of Rugby Park in Kilmarnock with a scale factor of 0·0004.

(a) Write down the area factor of the reduction.

(b) If the area of the pitch at Rugby Park is 7102 m², calculate the area of the scale drawing.

(c) The length of the pitch at Rugby Park is 106 m. Make a scale drawing of the pitch with scale factor 0·004.

B3 Semi-circular hearthrugs come in three sizes.

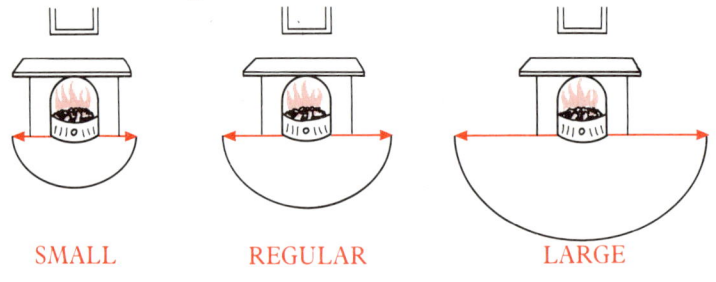

SMALL   REGULAR   LARGE

(a) The diameter of the small rug is 0·75 times the diameter of the regular rug.
How many times smaller is the area of the small rug than the area of the regular rug?

(b) The area of the large rug is 2·25 times the area of the regular rug. How much smaller is the diameter of the regular rug than that of the large rug?

(c) If the small rug has diameter 1·5 m and the large rug has area 3·53 m², calculate the diameter and area of the regular rug.

B4 Here is a pattern for a beach bag.

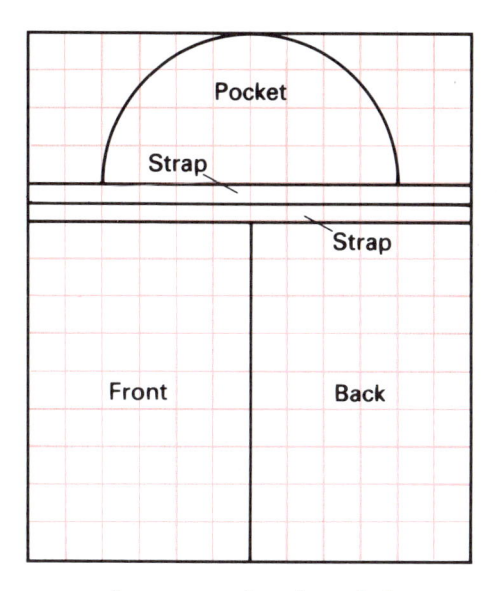

Every square on the pattern has length $\frac{1}{2}$ cm.
To make the bag, each square must be enlarged to length 5 cm.

(a) Write down the scale factor and area factor of the enlargement

(b) The instructions suggest that 0·7 m of 120 cm wide material is bought. How much material is wasted?

B5 Sheets of paper come in standard sizes A1, A2, A3, and so on. A sheet of A4 measures 297 mm by 210 mm.

The area of a sheet of A5 paper is $\frac{1}{2}$ that of a sheet of A4 paper which is $\frac{1}{2}$ that of a sheet of A3 paper, and so on.

(a) Work out the areas of an A5 sheet, an A3 sheet and an A1 sheet.

(b) What is the scale factor of the reduction from A4 to A5?

(c) Calculate the length of the diagonal of a sheet of A5 paper.

(d) Calculate the length of the diagonal of a sheet of A1 paper.

C Cubes

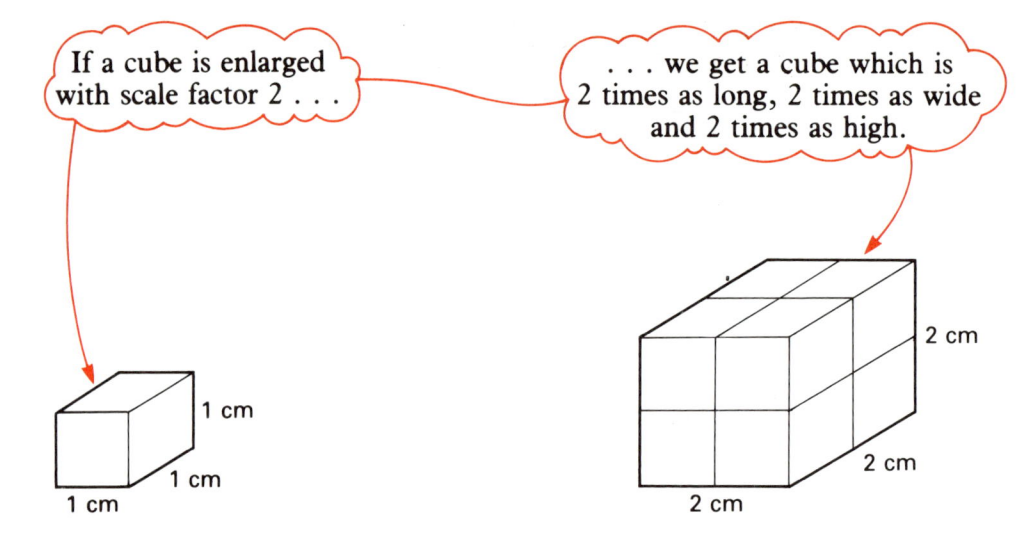

If a cube is enlarged with scale factor 2 . . .

. . . we get a cube which is 2 times as long, 2 times as wide and 2 times as high.

C1 (a) Write down the volume of each of the above cubes.

(b) By how many times has the volume of the smaller cube been enlarged?

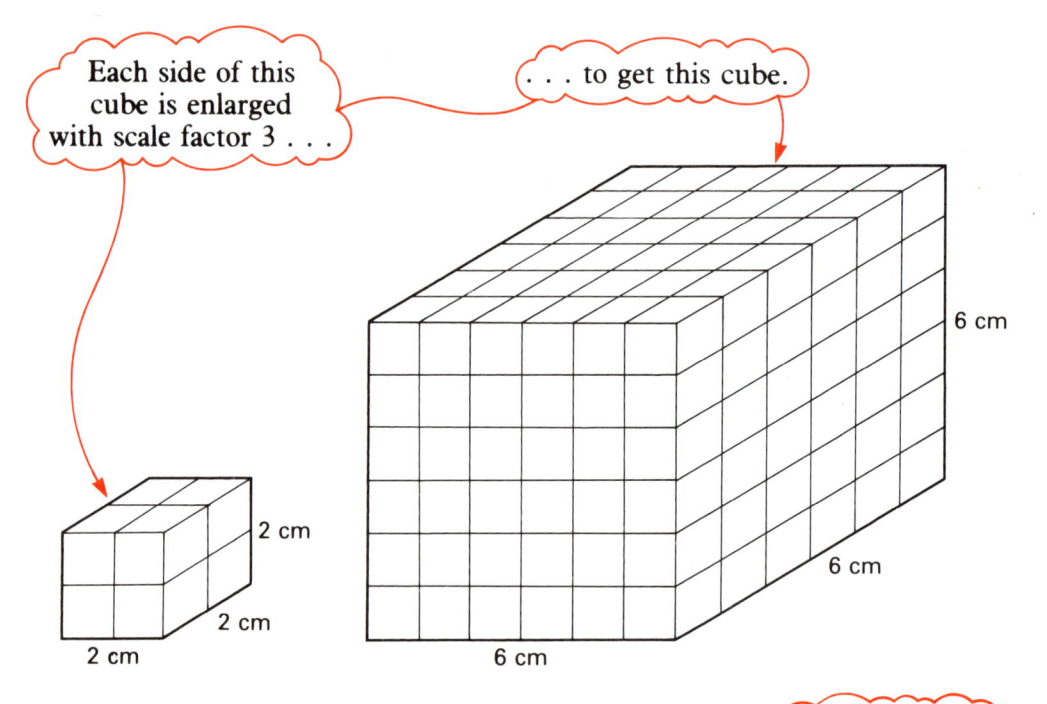

Each side of this cube is enlarged with scale factor 3 . . .

. . . to get this cube.

C2 (a) Write down the volumes of these cubes.

*This is called the **volume factor**.*

(b) Copy and complete:
The volume of the larger cube is ___ times the volume of the smaller cube.

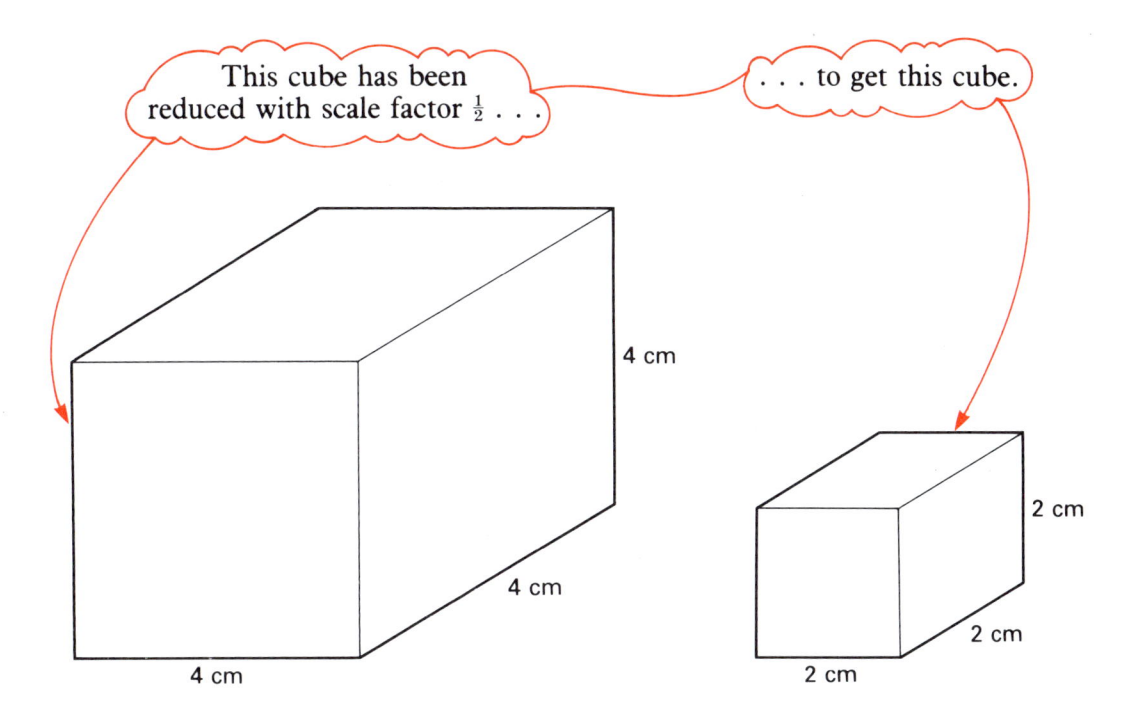

This cube has been reduced with scale factor ½ ...

... to get this cube.

C3 (a) Write down the volumes of these cubes.

(b) Copy and complete:
The volume of the smaller cube is ___ times the volume of the larger cube.

C4 Write down the volume factor for each pair of cubes.

(a)

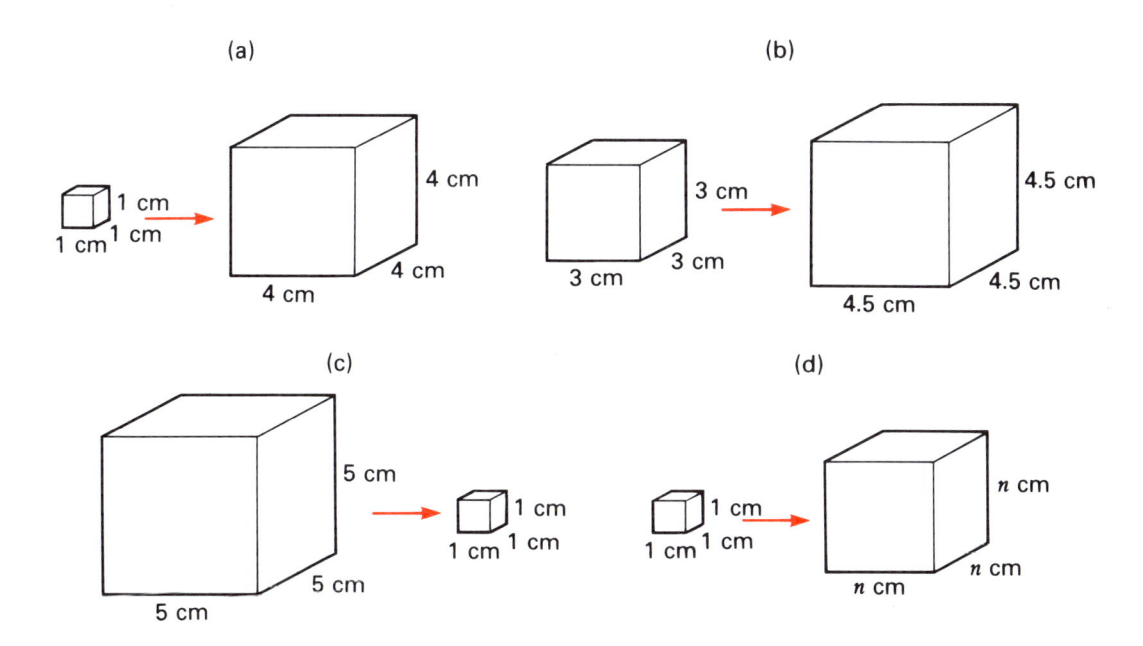

(b)

(c)

(d)

55

C5 Copy and complete this table.

Scale factor	Volume factor
3	
	125
0·2	
	$\frac{1}{8}$
$\frac{1}{3}$	

D Other volumes

So far we have found that when enlarging or reducing cubes.

- if the scale factor is 2, the volume factor is 8, or 2^3
- if the scale factor is $\frac{1}{3}$, the volume factor is $\frac{1}{27}$, or $(\frac{1}{3})^3$

In general we can say

Volume factor = (Scale factor)³

This rule is true whatever volume is enlarged or reduced.

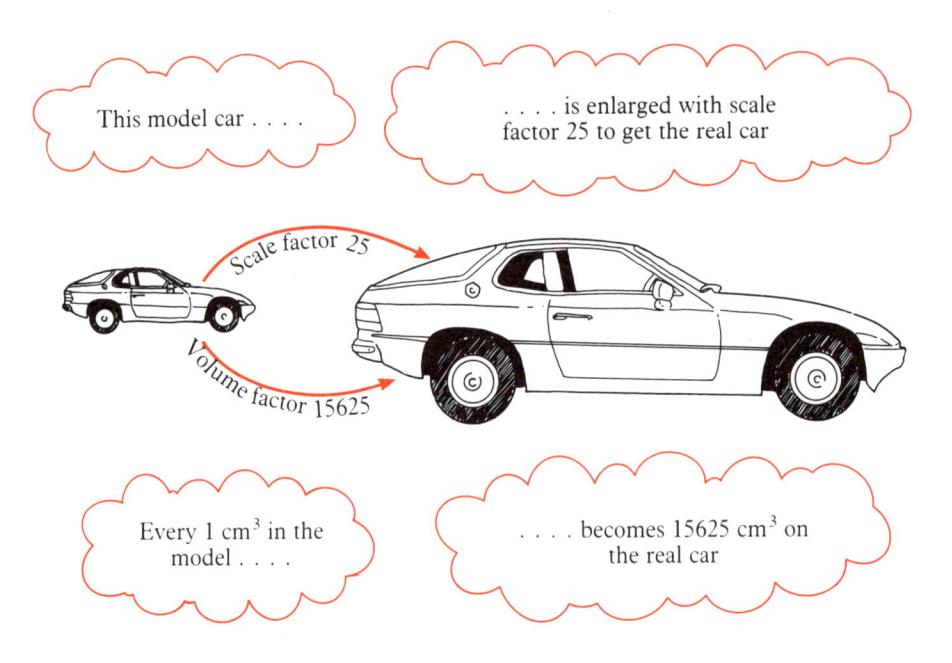

This model car . . . .

. . . . is enlarged with scale factor 25 to get the real car

Scale factor 25

Volume factor 15625

Every 1 cm³ in the model . . . .

. . . . becomes 15625 cm³ on the real car

D1 If the fuel tank of the model could hold 3 ml of fuel, how much does the fuel tank of the real car hold?

D2 This spoon holds 10 ml of liquid. It is enlarged with scale factor 1·5.

How much liquid will the enlarged spoon hold?

D3 This pan holds 1 litre.

This pan holds 2 litres.

Find the scale factor of the enlargement correct to 2 decimal places.

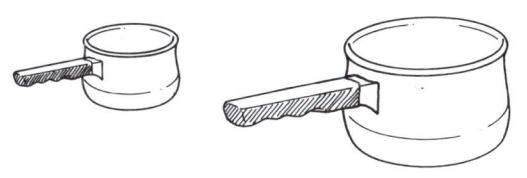

D4 In this tank, the young fish is 0·6 times the length of its mother. If the small fish weighs 5 g, how much would you expect the adult fish to weigh?

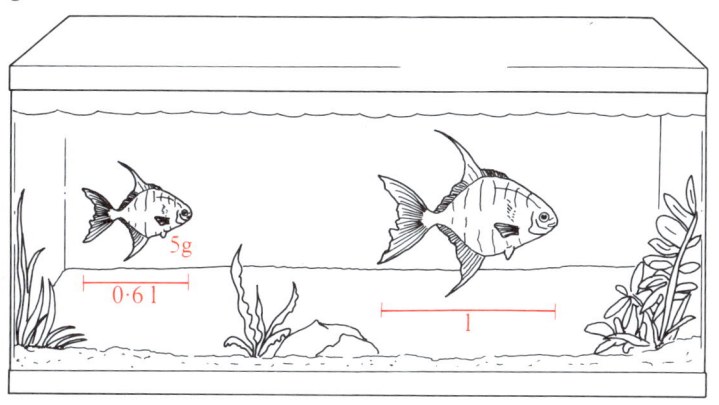

E Some similar problems

E1 Here are two bars of chocolate.

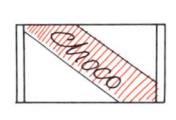

Neopolitan

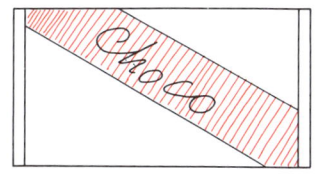

The neapolitan is approximately 0·25 times the size of the regular bar.

(a) Calculate the volume factor of the reduction.

(b) If the regular bar weighs 50 g, how much would you expect the neapolitan to weigh?

(c) If the weight of the neapolitan is increased to 1 g, by how much should the weight of the regular bar increase?

(d) The wrapper of the neapolitan uses 5 cm² of paper. How much paper is needed to make a wrapper for the regular bar?

E2 Here is a picture of a scale model of a 'Douglas Dauntless' aircraft.

The model is $\frac{1}{72}$ full size.

(a) Copy and complete the following statements about the model and the real plane.

1 cm on the model represents ___ cm on the real plane.
1 cm² on the model represents ___ cm² on the real plane.
1 cm³ on the model represents ___ cm³ on the real plane.

(b) This table gives some information about the model and the real plane. Copy it and fill in the blanks.

	Model	Real
Length	137 mm	
Wingspan		12·6 m
Area of wing	5200 mm²	

(c) The stalling speed of the plane is proportional to the square root of its length. For example, if the length is quartered then the stalling speed halves.

What fraction of the stalling speed of the real plane would be the stalling speed of the model?

(d) The plane's ability to fly depends on the ratio

$$\frac{\text{area of wings}}{\text{weight of plane}}$$

Would the model fly better or worse than the real plane?

E3 In the fairy tale 'Jack and the Beanstalk', Jack meets a giant 3 times the height of a normal man.

(a) How many times the weight of a normal man would the giant be?

(b) How many times bigger would the area of the cross-section of the giant's bones be than that of a normal man?

(c) How many times the weight borne by 1 cm² of human bone would 1 cm² of giant bone bear?

(d) Would the giant find it easier or harder to walk than a normal man?

E4 Yerother National Ginger plc make three sizes of bottle for their product 'Girders'.

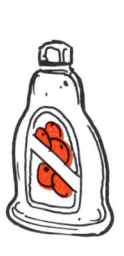

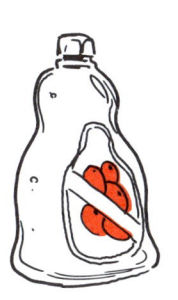

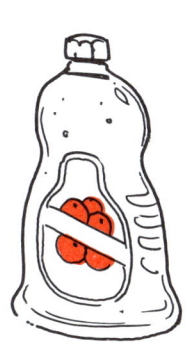

The Rusty and Roaster bottles are enlargements of the Rivet.

The Rivet bottle is 10 cm high and contains 250 ml of Girders.
The Rusty bottle is 15 cm high.

(a) Write down the scale factor of the enlargement.

(b) Calculate the volume factor and work out how much Girders the Rusty bottle holds.

The Roaster bottle contains 1 litre of Girders.

(c) Calculate the height of the Roaster bottle.

(d) The Rivet costs 32p, the Rusty costs 90p and the Roaster costs £1·50. Which bottle gives best value for money?

E5 Gleamo washing powder is packed in boxes of standard sizes E1, E2, E3 and E10. The E2 packet holds twice as much as the E1 packet and E3 packet holds three times as much as the E1 and so on.

(a) If an E3 packet holds 1155 g, how much would you expect an E10 pack to hold?

(b) The E3 pack costs £1·09 and an E10 packet costs £3·48. Which pack gives better value for money?

(c) The E3 pack measures 160 mm by 230 mm by 60 mm. What should be the dimensions of an E10 pack?

E6 The weight of an adult elephant is about 8 times that of an elephant calf.

How many times bigger will the large elephant be in length?

F Areas and volumes in nature

The ratio $\frac{\text{total surface area}}{\text{volume}}$ is very important in nature, as the following examples show.

It would be very difficult to measure accurately the surface area of a living animal, so in the examples we have used approximations only, ignoring small parts such as tail, ears, legs, etc.

F1 Here is a model of a young dog. As it grows, it increases in length and breadth and height.

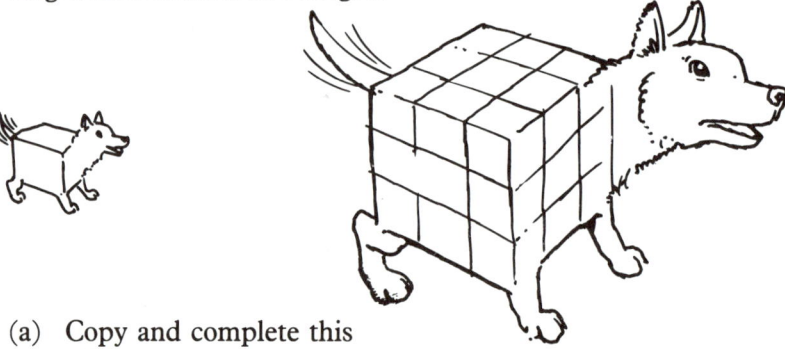

(a) Copy and complete this table.

Length	Surface area	Volume	Surface area / Volume
1 unit			
2 units			
3 units			

(b) What happens to the $\frac{\text{surface area}}{\text{volume}}$ ratio as the dog grows?

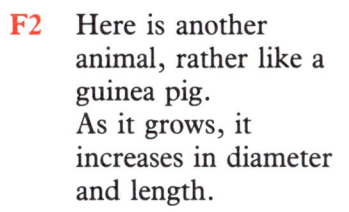

F2 Here is another animal, rather like a guinea pig. As it grows, it increases in diameter and length.

(a) Copy and complete this table.

Length	Diameter	Surface area	Volume	$\frac{\text{Surface area}}{\text{Volume}}$
2 units	1 unit			
4 units	2 units			
6 units	3 units			

(b) What happens to the $\frac{\text{surface area}}{\text{volume}}$ ratio as the animal gets bigger?

F3 In general, what happens to the $\frac{\text{surface area}}{\text{volume}}$ ratio as animals get bigger?

Smaller animals have a larger surface area in comparison with their volume than larger animals. This means that they will lose heat more rapidly than larger animals, as the following examples show.

F4 These three flasks have volumes 500 ml, 100 ml and 50 ml. They are all heated to the same temperature, 100 °C, and then the heat is switched off.

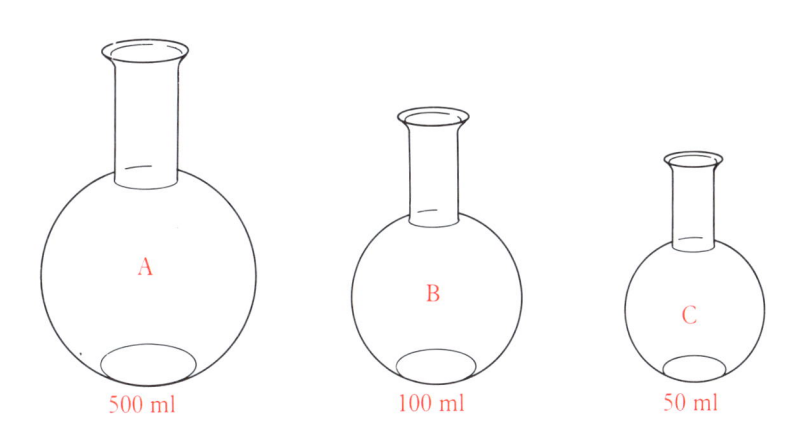

A 500 ml

B 100 ml

C 50 ml

The temperature of each flask is measured every 5 minutes. The results are shown in the table.

	Flask	\multicolumn{7}{c}{Time in minutes}
Temperature		0	5	10	15	20	25	30
	A	100	86	70	60	52	45	38
	B	100	66	50	38	30	23	18
	C	100	58	42	30	20	12	8

(a) Draw the graphs of (time, temperature) for the three flasks on the same grid.

(b) Copy and complete this table.

Flask	A	B	C
Radius	5 cm	3 cm	2·5 cm
Approx. surface area			
Volume	500 ml	100 ml	50 ml
Surface area / Volume			

For a sphere, radius r, surface area $= 4\pi r^2$

(c) Discuss the $\frac{\text{surface area}}{\text{volume}}$ ratios with reference to the graphs.

(d) If three animals of the same species, similar in sizes to these flasks, were exposed to extreme cold, which would suffer first? Which would suffer least?

In nature, small changes are made to the sizes and shape of animals within a species to make the $\frac{\text{surface area}}{\text{volume}}$ ratio smaller in those animals who are exposed to the coldest conditions. This means that they lose heat less rapidly.

F5 Look at these different types of fox.

Which do you think lives in the coldest conditions?

F6 Here is some information about different types of penguins.

Hint. Use the area factor.

The weight of an animal is directly related to its volume

Name	Height	Surface area	Weight
Emperor	100 cm	x cm²	40 kg
Galapagos	50 cm		20 kg
King	80 cm		50 kg

By considering the $\dfrac{\text{surface area}}{\text{volume}}$ ratio, arrange the penguins in the order you would expect to find them if you were travelling from the equator to the south pole.

G Investigation

Investigate how the $\dfrac{\text{surface area}}{\text{volume}}$ ratio influences

(a) water loss in objects of differing size

(b) the ability of a bird to fly

(c) how well an animal can resist a fall

Money management 7: Energy bills

Now Cathy and Colin are living in their own home, they find that they have several regular bills to pay. One of these is for electricity. It has to be paid every two months.

A Units of electricity

Different electrical appliances use up differing amounts of electricity. This is given by their **wattage**.

Wattages are given in **watts** (W) or **kilowatts** (kW). 1000 W = 1 kW

One unit of electricity is the amount used up by a 1 kW appliance in one hour.

Here are some examples.

These curling tongs are 200 W.
They use up one unit of electricity in 5 hours.

This fan heater is 2 kW.
It can run for $\frac{1}{2}$ hour on one unit.

A1 How long do the following appliances run on one unit of electricity?

(a) Microwave oven (500 W)   (b) Kettle (2 kW)   (c) TV (125 W)

A2 Copy and complete this table.

Appliance	Wattage	Units used in 1 hour
Electric blanket	50 W	
Hair dryer		$\frac{1}{2}$
Electric fire		3
Fluorescent light		0·05

A3 Complete these:

(a) 8 hours of 125W costs the same as ___ hours of 500W

(b) 1 hour of 2 kW costs the same as ___ hours of 250W

(c) 5 hours of 400W costs the same as ___ hours of 500W

A4 This cartoon shows a typical Saturday in Cathy and Colin's life. Work out the total amount of electricity used in the cartoon.

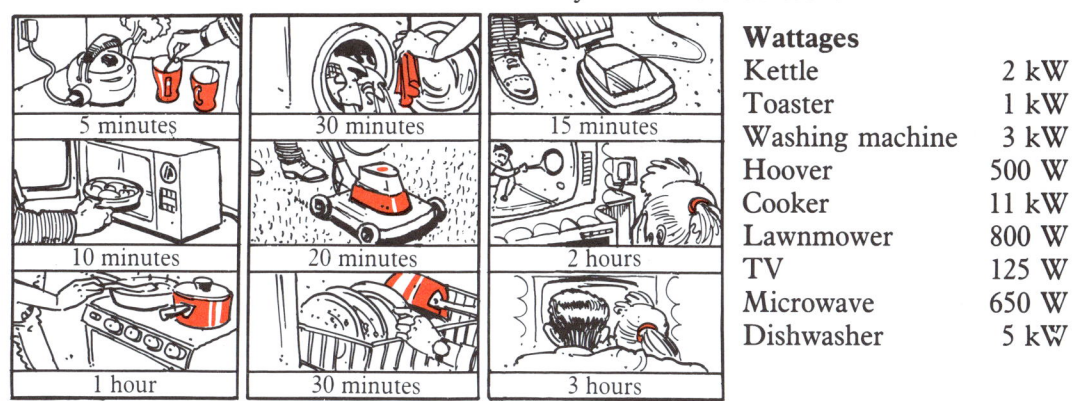

Panel	Time
(Kettle/drinks)	5 minutes
(Washing machine)	30 minutes
(Toaster)	15 minutes
(Microwave)	10 minutes
(Hoover)	20 minutes
(TV)	2 hours
(Cooker)	1 hour
(Dishwasher)	30 minutes
(Hairdryer)	3 hours

Wattages

Appliance	Wattage
Kettle	2 kW
Toaster	1 kW
Washing machine	3 kW
Hoover	500 W
Cooker	11 kW
Lawnmower	800 W
TV	125 W
Microwave	650 W
Dishwasher	5 kW

B The cost of electricity

Each unit of electricity costs Cathy and Colin 5·2p (in 1989).

B1 Calculate the cost of running each of the appliances in question A2 for an hour.

The South of Scotland Electricity Board gives its customers a 'ready reckoner' like this to work out the approximate cost of their electricity.

DOMESTIC TARIFF   5.20p per unit

Units	£	p	Units	£	p	Units	£	p	Units	£	p
1		5	10		52	100	5	20	1000	52	00
2		10	20	1	04	200	10	40	2000	104	00
3		16	30	1	56	300	15	60	3000	156	00
4		21	40	2	08	400	20	80	4000	208	00
5		26	50	2	60	500	26	00	5000	260	00
6		31	60	3	12	600	31	20	6000	312	00
7		36	70	3	64	700	36	40	7000	364	00
8		42	80	4	16	800	41	60	8000	416	00
9		47	90	4	68	900	46	80	9000	468	00

A Standing Charge bi-monthly under this tariff is £4.04 and is payable whether or not electricity is consumed.

B2 Use the ready reckoner to work out the cost of the following amounts of electricity.

(a) 80 units (b) 300 units (c) 70 000 units
(d) 58 units (e) 590 units (f) 4325 units

B3 Work out the cost of running the following

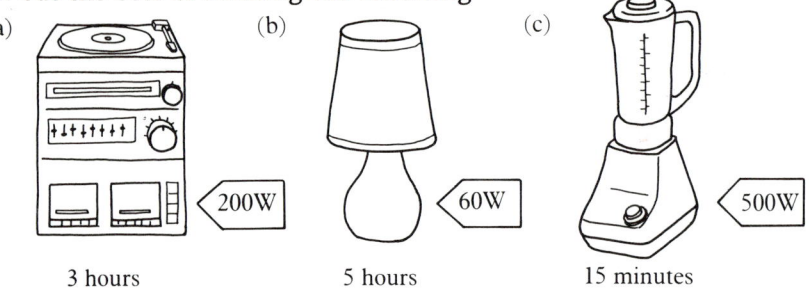

(a) 200W, 3 hours

(b) 60W, 5 hours

(c) 500W, 15 minutes

B4 Try to work out how much electricity you use in a day. How much does it cost?

C The bill

Working out the amount of electricity used in a household would be very difficult and time consuming, if you had to do it as in question B4. Instead, the amount used is measured by a meter like this.

This reading shows 13569·5 units. | 1 | 3 | 5 | 6 | 9 | 5 |

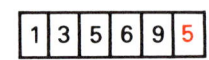

C1 Calculate the amount of electricity used and its cost between these pairs of meter readings.

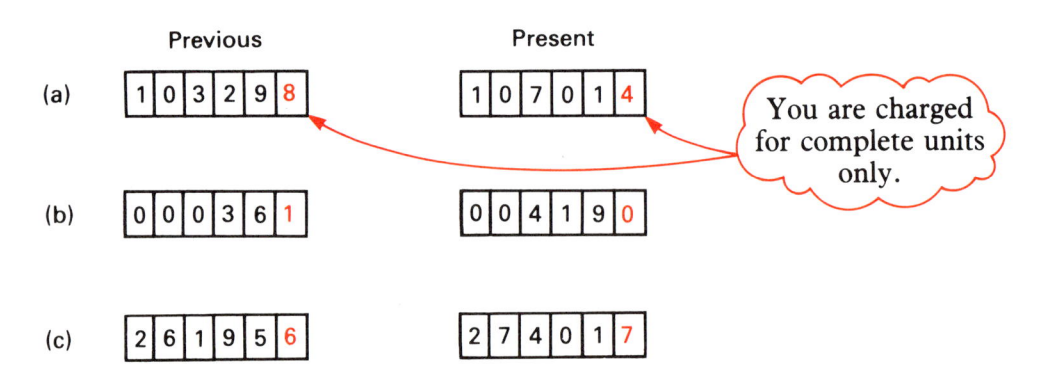

Previous   Present

(a) 1 0 3 2 9 8   1 0 7 0 1 4   *You are charged for complete units only.*

(b) 0 0 0 3 6 1   0 0 4 1 9 0

(c) 2 6 1 9 5 6   2 7 4 0 1 7

Apart from the cost of the electricity used, the Electricity Board adds on a set amount to each bill. This amount must be paid even if no electricity is used. It is called a **standing charge**.

The charge for Cathy and Colin is £2.02 per month.

Here is an example.

C2 Copy and complete this electricity bill.

1ST MARCH – 30TH APRIL

| METER READING || DETAILS | AMOUNT |
PRESENT	PREVIOUS		
34290	31452	? @ 5.2p	?
		STANDING CHARGE	?
		TOTAL DUE	?

C3 Copy and complete these bills.

1ST MAY – 30TH JUNE

(a)

| METER READING || DETAILS | AMOUNT |
PRESENT	PREVIOUS		
00981	?	502 @ 5.2p	?
		STANDING CHARGE	£4.04
		TOTAL	?

1ST NOVEMBER – 31ST DECEMBER

(b)

| METER READING || DETAILS | AMOUNT |
PRESENT	PREVIOUS		
39012	?	? @ 5.2p	£111.28
		STANDING CHARGE	£4.04
		TOTAL	?

7 Circles

A Sectors

A sector of a circle is rather like a slice of cake.

A1 Mary is given a slice of cake. What fraction of the cake is this?

A2 Here is a circle with a sector marked.

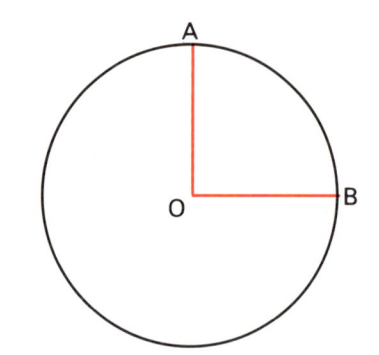

(a) What fraction of the circle is this?

(b) What size is angle AOB?

(c) If the circumference of the circle is 12·6 cm, what length do you think arc AB will be?

(d) Check your answers to (b) and (c) by measuring.

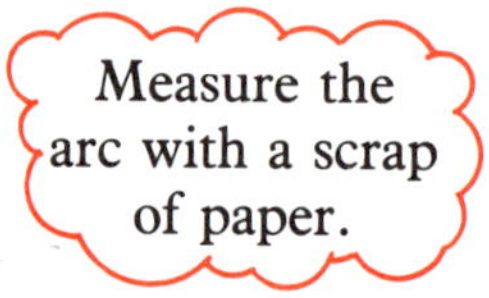

Measure the arc with a scrap of paper.

A3 Here is another circle with a sector marked.

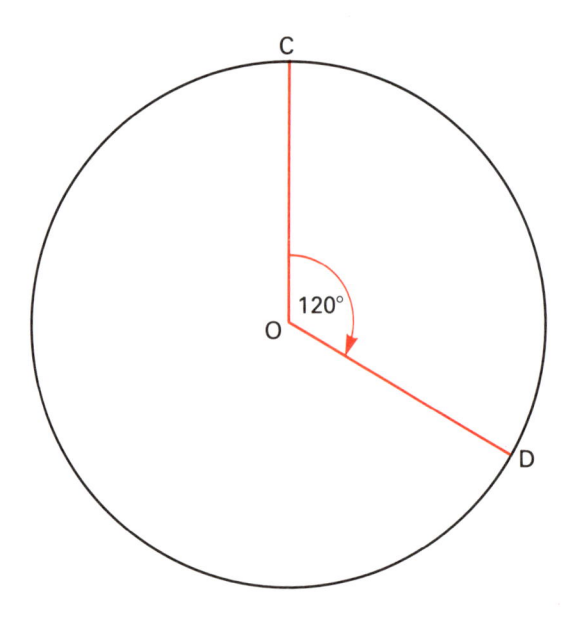

(a) What fraction of the circle is this sector?

(b) The area of the circle is 34·48 cm². What is the area of sector COD?

(c) If the arc CD measures 7·33 cm, what is the circumference of the circle?

A4 (a) Work out the circumference and area of this circle.

(b) What fraction of the circle is sector XOY?

(c) What is the length of arc XY?

(d) What is the area of sector XOY?

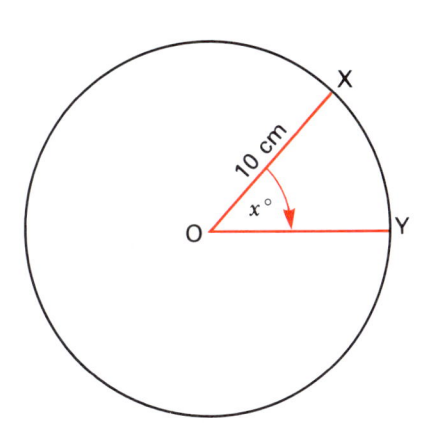

A5 Repeat question A4 for these two diagrams.

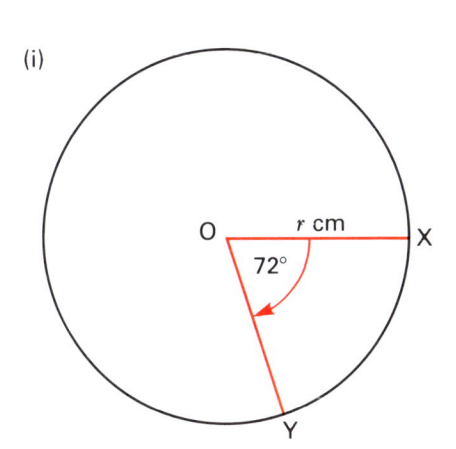

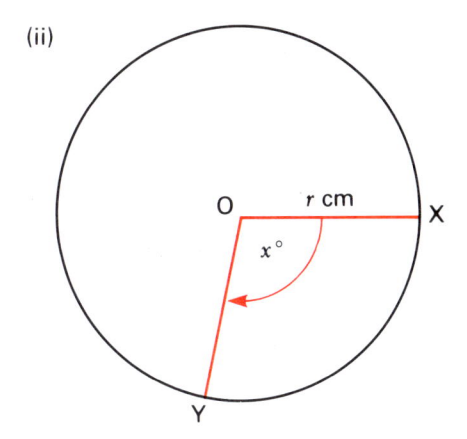

(i)

(ii)

A6 The area of the shaded sector is 2·512 cm². Calculate the area of the circle.

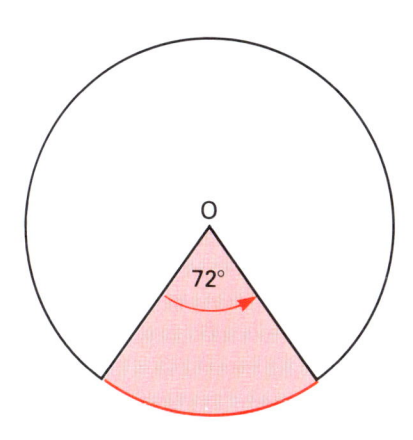

69

A7 The length of arc AB is 12·56 cm.
Calculate the circumference of the circle.

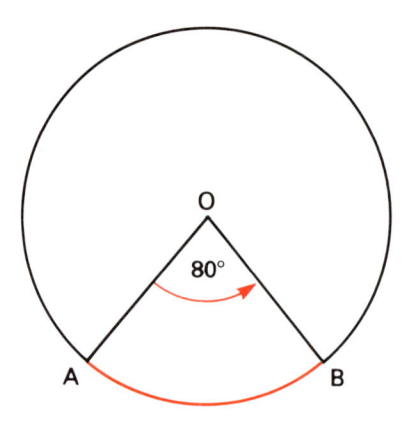

A8 This sector is to be formed into a cone.

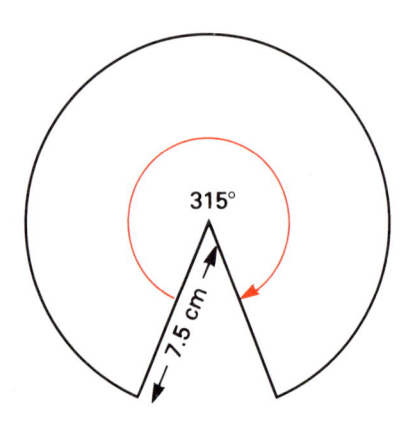

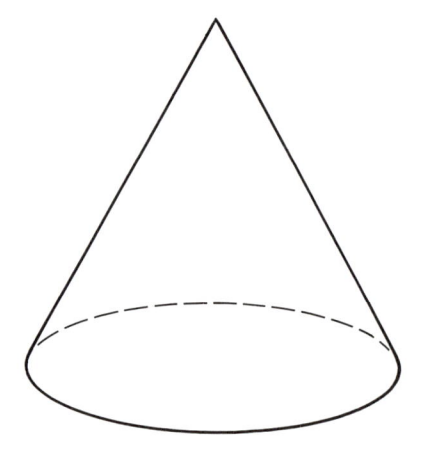

(a) Work out the circumference of the cone.

(b) Work out the vertical height of the cone.

A9 The length of arc AB is 2π cm.

Calculate

(a) the circumference
(b) the radius
(c) the area

of the circle, leaving your answers in terms of π.

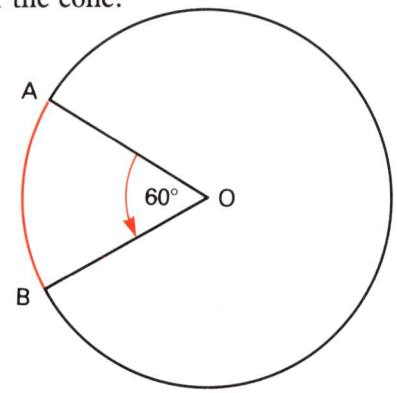

A10 The area of the shaded sector is 175π cm².

(a) Calculate the area of the circle, in terms of π.

(b) Calculate the radius of the circle, in terms of π.

(c) Calculate the length of arc AB, in terms of π.

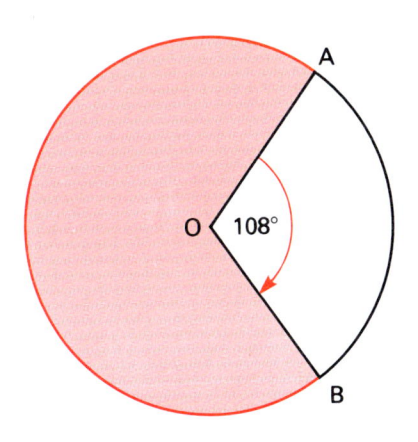

A11 These two circles have equal diameters.

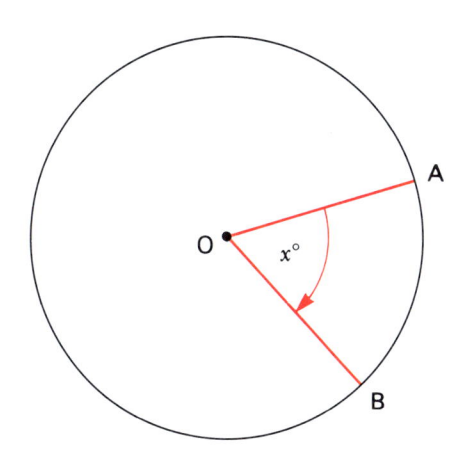

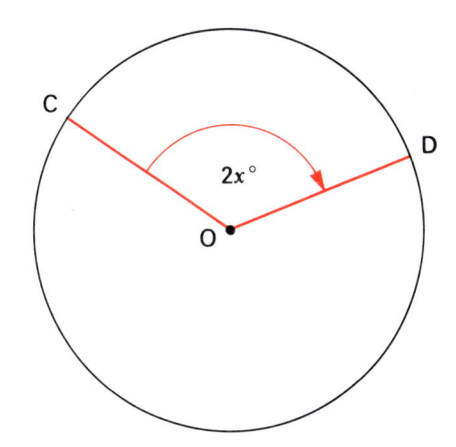

(a) What can you say about the lengths of arcs CD and AB?

(b) What can you say about the areas of sectors AOB and COD?

(c) The area of sector COD is 100 cm². Work out an expression for the area of the circle.

A12 These two circles have equal radii, *r* cm.

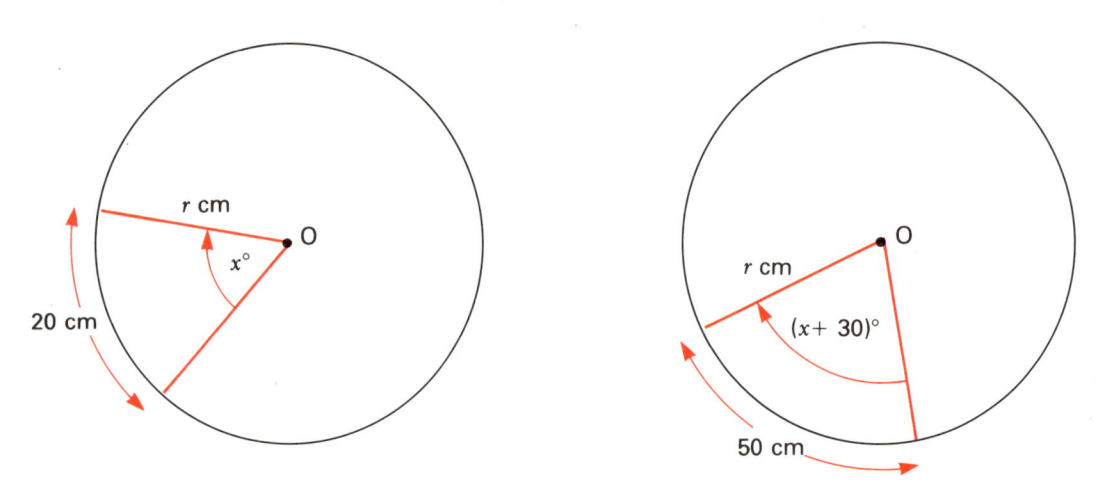

(a) Find the angles $x°$ and $(x + 30)°$.

(b) Find the radius of the circles.

B Tangents

A **tangent** is a line which touches a circle at one point only.

Here are some examples.

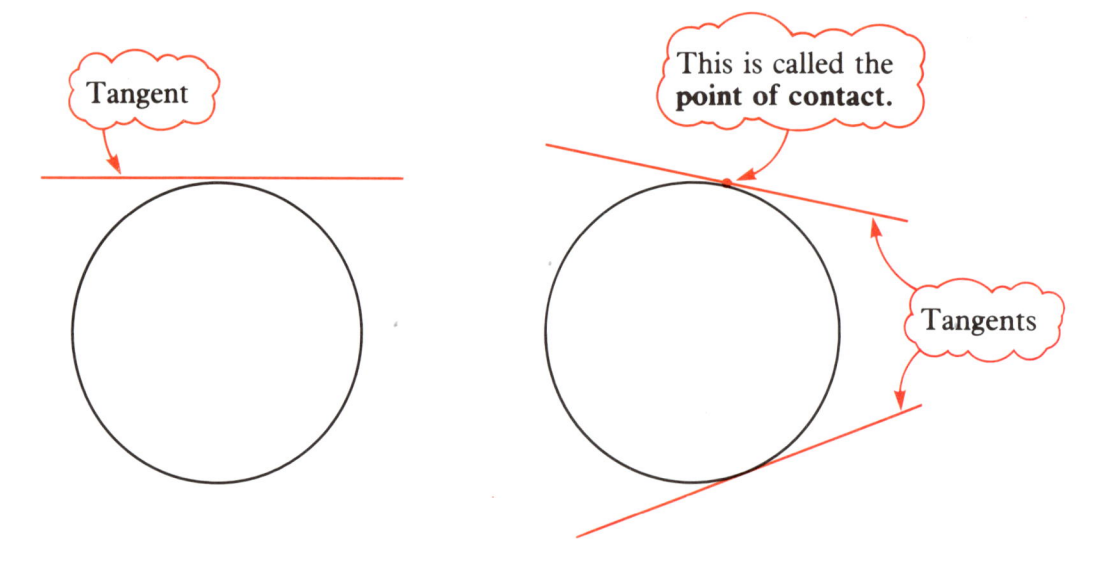

We shall now try to discover some properties of tangents.

B1 (a) Draw a circle and one of its tangents.

Use a ruler to join the centre of the circle to the point of contact.

(b) Measure the marked angle. Write down its size.

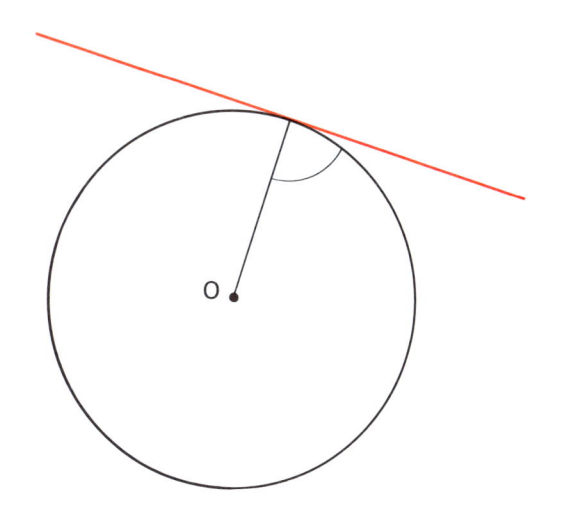

B2 Repeat question B1 for two more tangents.

B3 Repeat question B1 for two more circles.

B4 Write down your findings as a general rule.

B5 Draw a circle and mark any point outside it.

(a) How many tangents can you draw from the point to the circle?

(b) Draw these tangents. Measure the distance from the point to the points of contact. What do you notice?

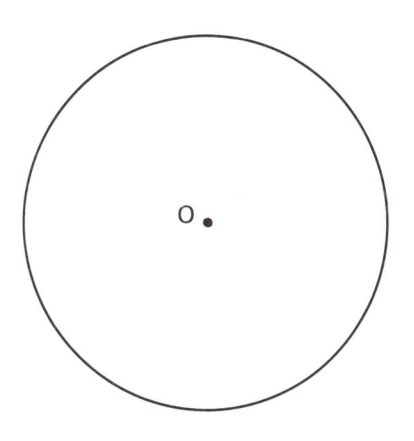

B6 Repeat question B5 for two different circles.

B7 Write down your findings as a general rule.

B8 Copy this diagram.

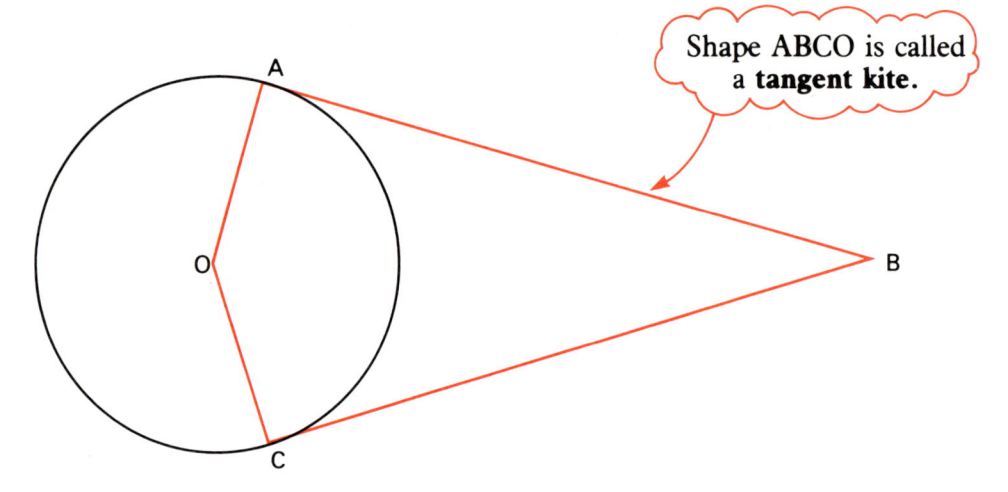

Shape ABCO is called a **tangent kite**.

(a) Why do you think the shape is called a tangent kite?

(b) Write down some properties of a tangent kite. Mark them clearly on your diagram.

B9 AP and BP are tangents to the circle with centre O.

Write down a relationship between x and y.

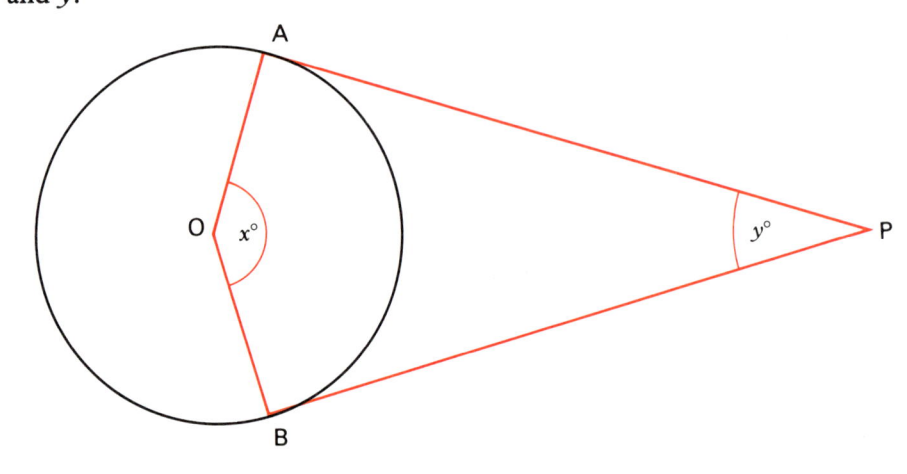

B10 AB, AC and BC are tangents to the circle, touching at X, Y and Z respectively.

Calculate the sizes of angles A, B and C.

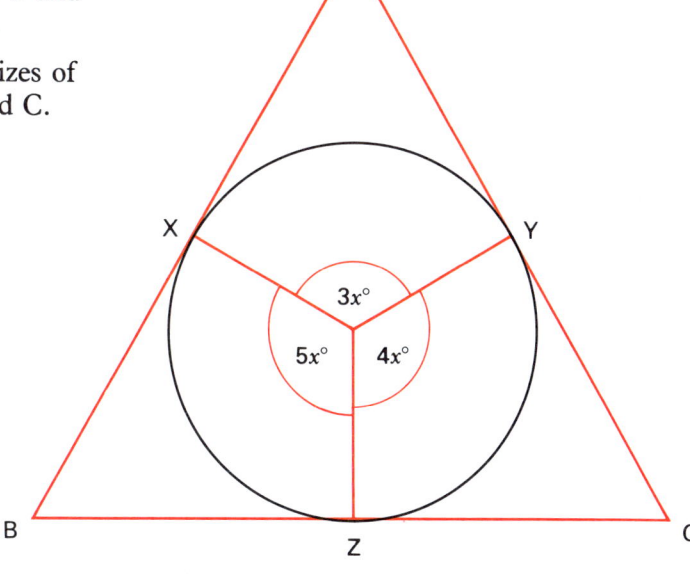

C Properties of angles

C1 (a) Draw a circle and mark on one diameter. Mark one other point on the circumference. Join each end of the diameter to this point. Measure the angle formed and write down its size.

(b) Try this again with the same diameter but with two different points on the circumference. What do you notice?

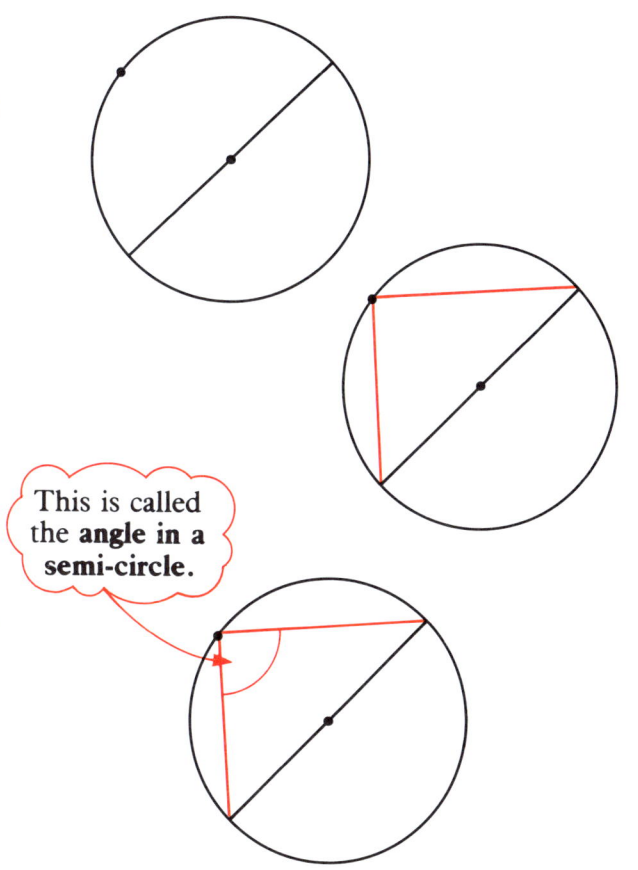

This is called the **angle in a semi-circle**.

75

C2 Repeat question C1 for two more circles.

C3 Why do you think this is called the angle in a semi-circle?

C4 (a) Draw a circle and mark two points, A and B, on the circumference.
Mark another point on the circumference, and join A and B to it. Measure the angle formed.

(b) Try this again using A and B with three other points on the circumference. What do you notice?

(c) Try this again for three more circles. What do you notice?

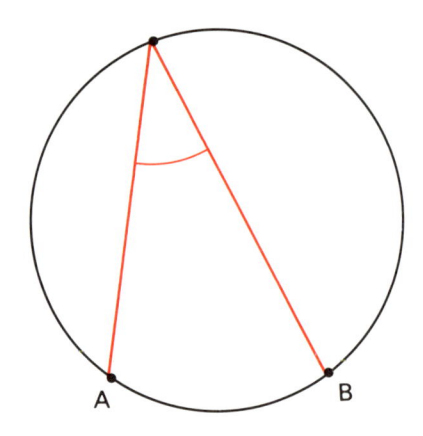

C5 (a) Draw a circle of any size and mark its centre, 0. Mark three points on the circumference and call them A, B and C

Join AC and BC. Measure angle ACB and write down its size.

(b) Join AO and BO. Measure angle AOB and write down its size.

(c) What do you notice about angles ACB and AOB?

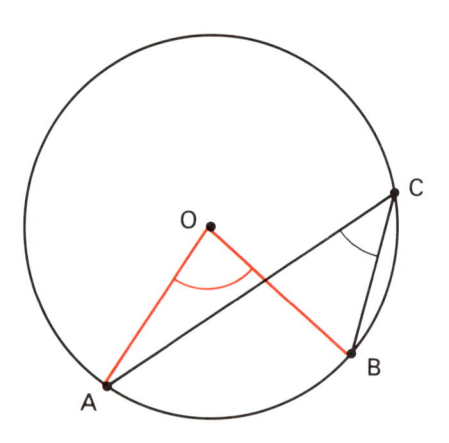

C6 Repeat question C5 for a few more circles. What do you notice?

C7 Work out the values of the angles marked **?** in these diagrams.

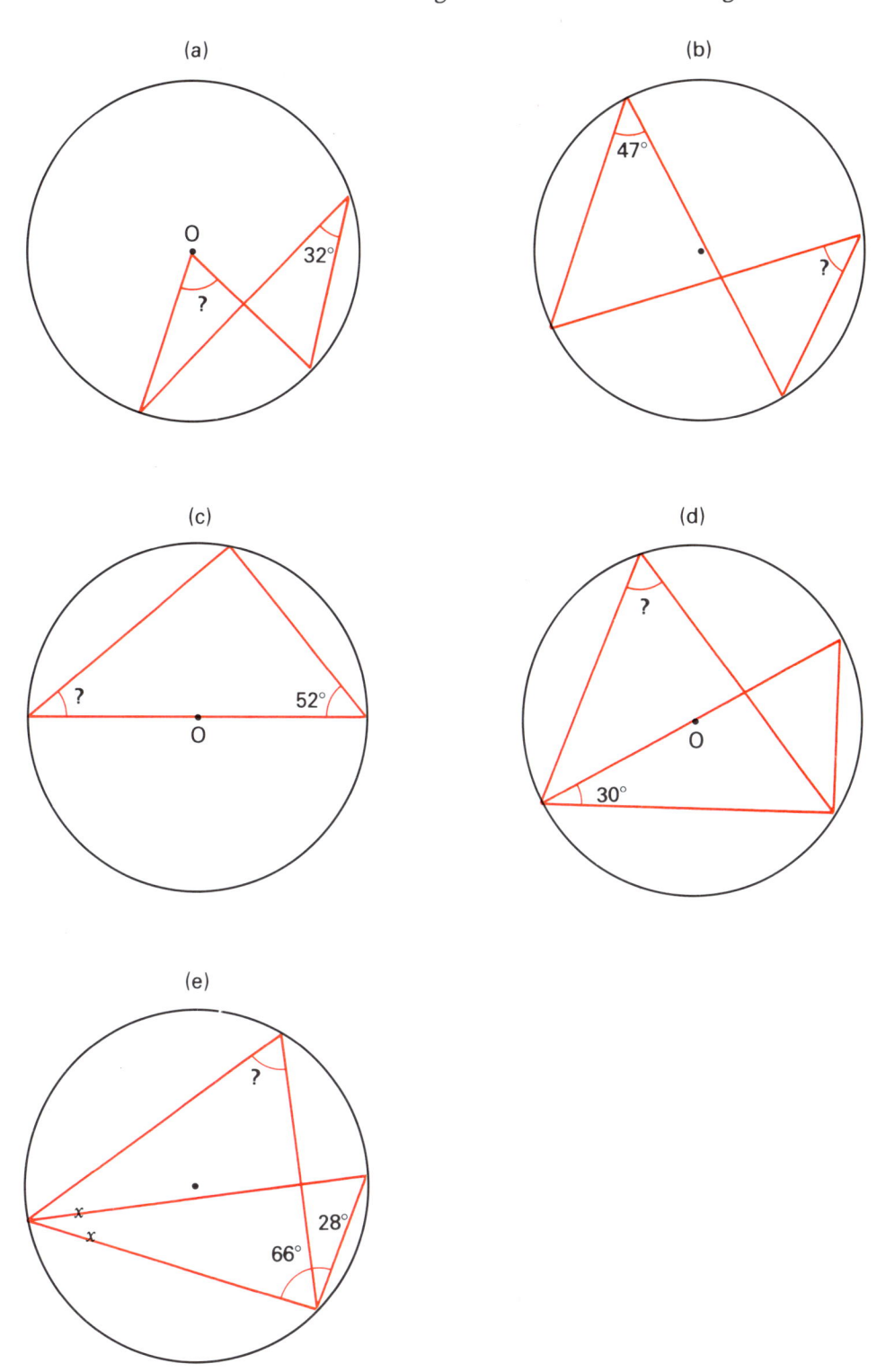

(a)

(b)

(c)

(d)

(e)

D Puzzles and problems

Mary has noted down all her findings from this chapter in her notebook. Check that you have noted the same properties.

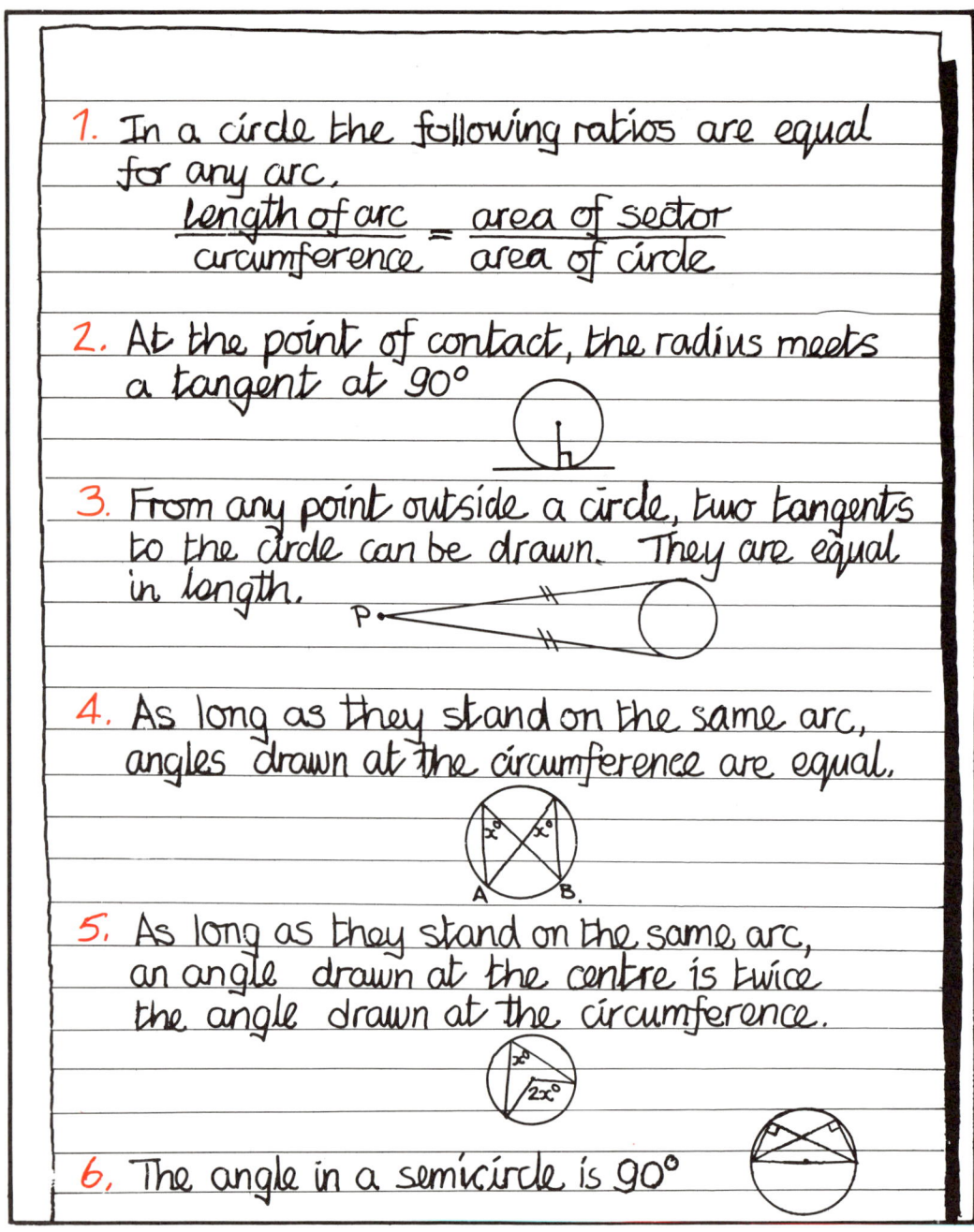

1. In a circle the following ratios are equal for any arc.
   $$\frac{\text{length of arc}}{\text{circumference}} = \frac{\text{area of sector}}{\text{area of circle}}$$

2. At the point of contact, the radius meets a tangent at 90°

3. From any point outside a circle, two tangents to the circle can be drawn. They are equal in length.

4. As long as they stand on the same arc, angles drawn at the circumference are equal.

5. As long as they stand on the same arc, an angle drawn at the centre is twice the angle drawn at the circumference.

6. The angle in a semicircle is 90°

Now use these properties to solve the following puzzles and problems.

Be careful! Sometimes you will have to look for clues to help you.

D1 Work out the sizes of the marked unknown angles in these diagrams.

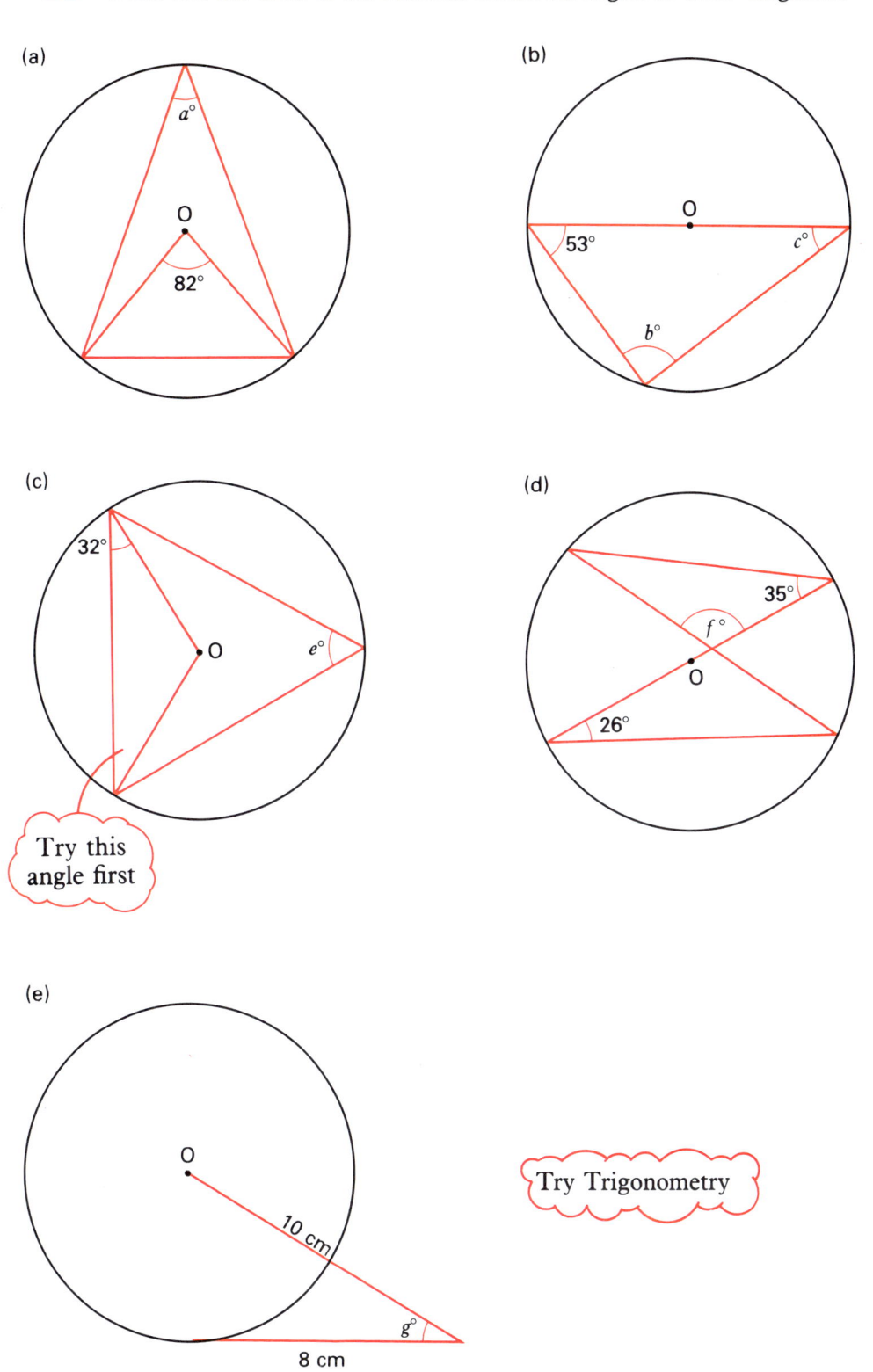

(a) 82° at O, a°

(b) 53°, b°, c°

(c) 32°, e°

(d) 35°, 26°, f°

Try this angle first

(e) 10 cm, 8 cm, g°

Try Trigonometry

D2 A satellite is in orbit 550 km above the Earth's surface directly above the north pole. From the spacecraft, point A on the Earth's surface can just be seen.

(a) If the radius of the Earth is 6400 km, work out the distance from the satellite to A to the nearest km.

(b) If the distance from the satellite to A is x km and the satellite is y km above the Earth, show that

$$x = \sqrt{y(12\ 800 + y)}$$

clearly explaining your reasoning.

D3 George is making a bench for his garden from part of an old tree trunk.

A cross-section of the bench looks like this.

(a) Calculate the radius of the trunk, to the nearest cm.

(b) If the length of the trunk is 1·5 m and the wood weighs 0·7 g per 1 cm³, calculate the weight of wood in the trunk.

(c) Calculate the weight of George's bench.

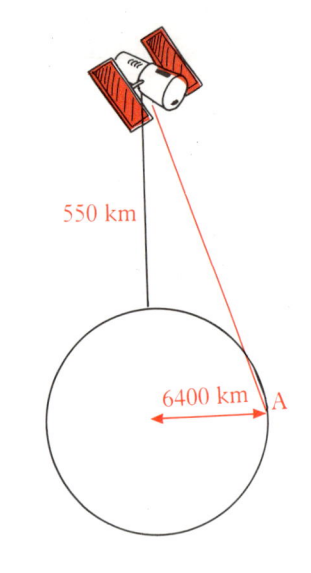

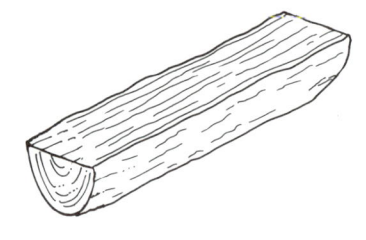

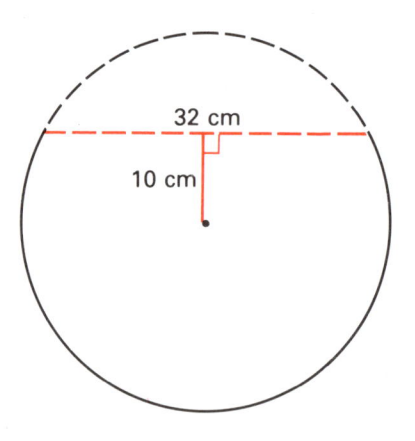

D4 The theorem of intersecting chords states that if any two chords AB and CD in a circle intersect at X then

$$AX \times BX = CX \times DX.$$

Draw a circle of any size. Show that this theorem works for any two intersecting chords drawn in your circle.

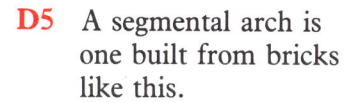

D5 A segmental arch is one built from bricks like this.

The two measurements which an architect needs to determine the curve of the arch are the span YZ and the rise RM.

(a) By extending RM to form a diameter and using the theorem of intersecting chords, determine the radius of the curve.

(b) Explain how considering the chords YR and RZ would help determine where the centre of the circle would be.

(c) If the bricks used are 15 cm long, make a scale drawing of the arch using a suitable scale.

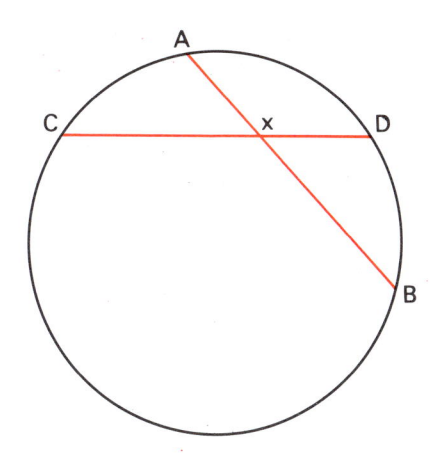

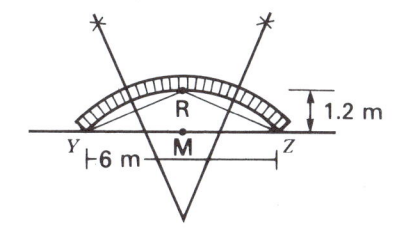

D6 'Dresden Plate' is a circular patchwork pattern which looks like this.

(a) What area of plain material would be required?

(b) Jane decides to make a Dresden Plate with 16 different prints instead of just two. How much of each material would she require?

'Grandmother's Fan' is a similar pattern which looks like this.

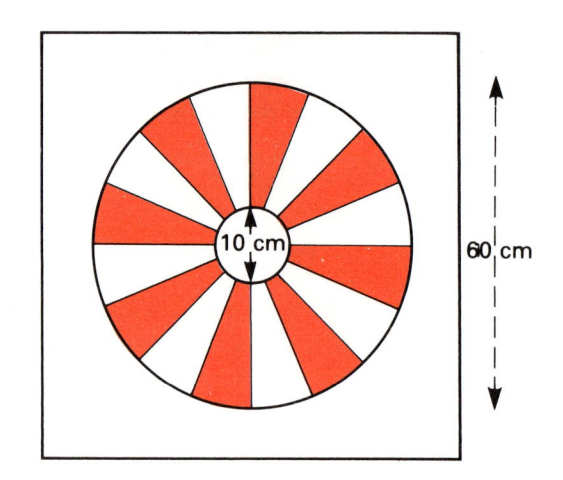

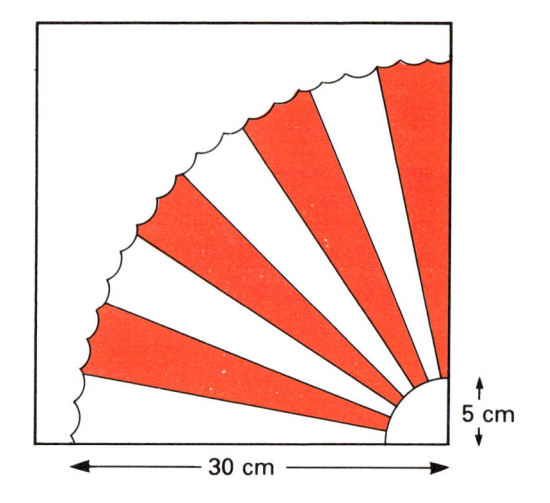

(c) Susan has made a Grandmother's Fan patch and wants to edge the fan with lace like this.

How much lace would she require?

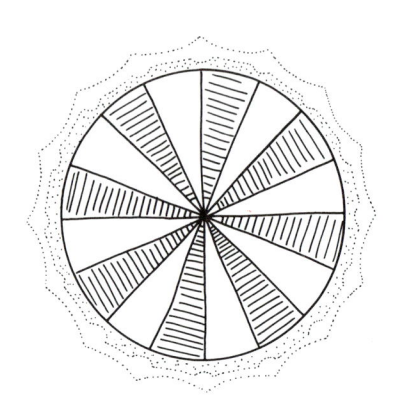

(d) Susan has made another Grandmother's Fan patch, but this time it is a bit smaller. If each wedge in the fan has an area of 3π cm², work out the radii of the quarter circles.

8 Algebraic fractions

A Cancelling down

Fractions like $\frac{10}{18}$ and $\frac{16}{24}$ are not in their *lowest terms*. They can be simplified by dividing the numerator and denominator by the same number.

$$\frac{10}{18} = \frac{5}{9} \quad \text{Dividing numerator and denominator by 2.}$$

$$\frac{16}{24} = \frac{2}{3} \quad \text{Dividing numerator and denominator by 8.}$$

Here is a different way to set them out.

$$\frac{10}{18} = \frac{\cancel{2} \times 5}{\cancel{2} \times 9} = \frac{5}{9} \qquad \frac{16}{24} = \frac{\cancel{8} \times 2}{\cancel{8} \times 3} = \frac{2}{3} \quad \text{This is called cancelling.}$$

Notice in both cases that there is a **common factor** in the numerator and the denominator which can be cancelled out.

Fractions involving letters like $\frac{9a^2b}{12ab}$ and $\frac{2(x+7)}{9(x+7)}$ can be simplified in the same way.

$$\frac{9a^2b}{12ab} = \frac{3a}{4} \quad \text{Dividing numerator and denominator by 3, } a \text{ and } b.$$

$$\frac{2(x+7)}{9(x+7)} = \frac{2}{9} \quad \text{Dividing numerator and denominator by } (x+7).$$

A1 Simplify the following by cancelling common factors.

(a) $\frac{2ab}{3ac}$ (b) $\frac{4bc}{c^2}$ (c) $\frac{3a^2}{6ab}$

(d) $\frac{24xy^3}{4x^2y^2}$ (e) $\frac{27ab^2c}{12a^3b}$ (f) $\frac{4(x-2)}{3(x-2)}$

(g) $\frac{(x+1)(x-4)}{(x-1)(x+1)}$ (h) $\frac{(2a-1)(a+1)}{(a-2)(a+1)}$ (i) $\frac{(a-1)^2(2a+1)}{(a+1)(a-1)}$

If either the numerator or denominator is not in a factorised form, then they must be factorised first.

Consider $\frac{x^2 - 4x - 5}{x^2 - 1}$

Factorising, we get $\frac{(x+1)(x-5)}{(x-1)(x+1)}$

Now cancelling as before $\dfrac{(x+1)(x-5)}{(x-1)(x+1)} = \dfrac{x-5}{x-1}$

A2 Simplify the following by factorising and cancelling.

(a) $\dfrac{3x+9}{x^2-9}$ (b) $\dfrac{x^2+x-6}{5x-10}$

(c) $\dfrac{x^2-5x+4}{x^2-2x-8}$ (d) $\dfrac{x^2-4x+3}{x^2-1}$

(e) $\dfrac{2x^2-18}{x^2-x-6}$ (f) $\dfrac{x^2-x-12}{2x^2-10x+8}$

(g) $\dfrac{3c^2+3c-6}{8c^2-8}$ (h) $\dfrac{4x^2-12x-40}{2x^2-50}$

(i) $\dfrac{4x^2-2x-6}{4x^2-9}$ (j) $\dfrac{3ax^2+3ax-18a}{6a^2x^2-6a^2x-12a^2}$

(k) $\dfrac{x^2+xy-2y^2}{x^2-y^2}$ (l) $\dfrac{a^2+4ab+3b^2}{a^2+2ab+b^2}$

B The 'negative' approach

These two students are arguing about how to simplify

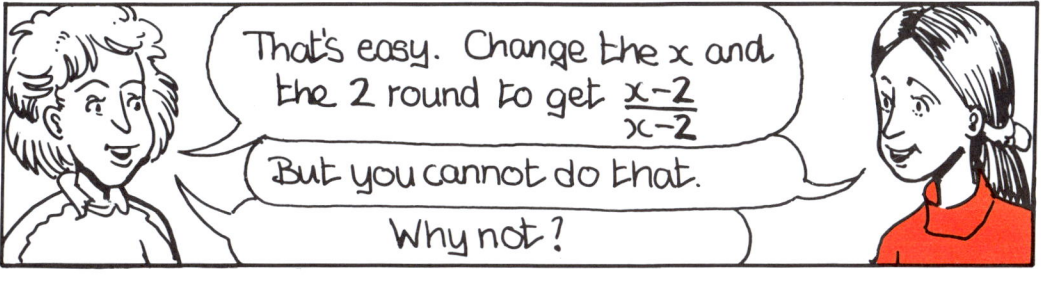

$\dfrac{x-2}{2-x}$

That's easy. Change the x and the 2 round to get $\dfrac{x-2}{x-2}$

But you cannot do that.

Why not?

B1 Why can you not just change the letter and the number round?

Try to find an example using numbers to show that this is not allowed.

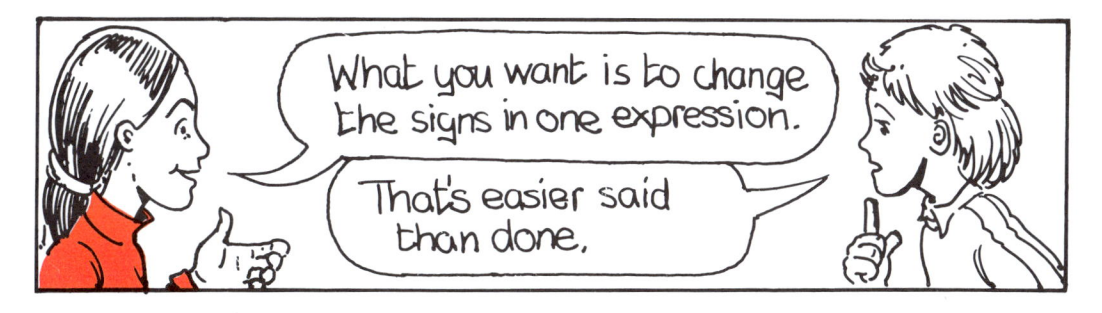

What you want is to change the signs in one expression.

That's easier said than done.

From SMP 11–16 *Book Y2*, you know

$-(a + b) = -a - b$ and $-(a - b) = -a + b$

B2 Remove the brackets in the following.

(a) $-(x - 2)$ (b) $-(7 + x)$ (c) $-(a - 5)$
(d) $-(-2 + b)$ (e) $-(-4 - d)$ (f) $-(p + 1)$
(g) $-(y - 6)$

B3 What do you notice about the terms inside the brackets and those outside the brackets in question B2?

So you remove the brackets to change signs...

But there aren't any brackets.

That's right. You have to put them in

Like doing question B2 backwards!

So take $x - 2$ and put in the brackets with a $-$ sign outside.

$x - 2 = -(\quad)$

So $x - 2 = -(-x + 2) = -(2 - x)$

This must be $+2$ since $-(+2) = -2$.

This must be $-x$ since $-(-x) = x$.

We can now simplify the fraction.

$$\frac{x - 2}{2 - x} = \frac{-(-x + 2)}{2 - x} = \frac{-(2 - x)}{2 - x}$$

$$= \frac{-1}{1}$$

$$= -1$$

Cancelling the $2 - x$ on numerator and denomintor.

*$-(2-x)$
$= -2+x$
$= x-2$*

B4 Copy and complete the following.

(a) $\dfrac{a}{b} = \dfrac{-a}{}$

(b) $\dfrac{a}{b} = -\dfrac{a}{}$

(c) $\dfrac{a}{a-b} = \dfrac{}{b-a}$

(d) $\dfrac{p-q}{q} = -\dfrac{p-q}{}$

(e) $\dfrac{x-y}{a-b} = \dfrac{y-x}{}$

(f) $\dfrac{x-y}{a-b} = -\dfrac{y-x}{}$

(g) $\dfrac{ab}{a-b} = \dfrac{-ab}{}$

(h) $\dfrac{p-q}{pq} = -\dfrac{}{pq}$

(i) $\dfrac{a(-b)}{a-b} = -\dfrac{ab}{}$

(j) $\dfrac{x+y}{x-y} = -\dfrac{x+y}{}$

(k) $\dfrac{1-x}{x} = -\dfrac{x-1}{}$

(l) $\dfrac{a}{a-b} = -\dfrac{-a}{}$

> Remember to take out a − sign if the signs are the wrong way round.

B5 Simplify the following.

(a) $\dfrac{2(x+y)}{y+x}$

(b) $\dfrac{5(a-b)}{b-a}$

(c) $\dfrac{a^2-b^2}{a^2+2ab+b^2}$

(d) $\dfrac{x^2-6x-7}{49-x^2}$

(e) $\dfrac{3-6x}{2x^2+x-1}$

(f) $\dfrac{8x+4}{1-4x^2}$

(g) $\dfrac{3x^2-12}{8-2x-x^2}$

(h) $\dfrac{5x^2+3x-2}{16-100x^2}$

(i) $\dfrac{8x^2+10x+3}{6+8x}$

C Fractions within fractions

$\dfrac{2}{3}$ is a fraction. So is $\dfrac{x-2}{7x+1}$

$\dfrac{\frac{2}{7}}{\frac{3}{8}}$ is a fraction within a fraction. So is $\dfrac{\frac{x+2}{2}}{\frac{3x-1}{x}}$

We know already that if we multiply the top and bottom of a fraction by the same number, its value remains the same. So to simplify $\dfrac{\frac{2}{7}}{\frac{3}{8}}$ we can multiply top and bottom by a number.

We choose this number so that the denominators of the fractions will both divide into it.

Multiplying by 56 we get

$$\dfrac{\frac{2}{7}}{\frac{3}{8}} = \dfrac{\frac{2}{7} \times \frac{56}{1}}{\frac{3}{8} \times \frac{56}{1}} = \dfrac{16}{21}$$

87

C1 Simplify these in the same way.

(a) $\dfrac{\frac{4}{3}}{\frac{1}{2}}$ (b) $\dfrac{\frac{3}{7}}{\frac{4}{3}}$ (c) $\dfrac{\frac{5}{12}}{\frac{1}{4}}$

(d) $\dfrac{\frac{3}{8}}{\frac{1}{6}}$ (e) $\dfrac{\frac{1}{12}}{\frac{6}{5}}$ (f) $\dfrac{\frac{3}{8}}{\frac{5}{4}}$

Take $\dfrac{\frac{x}{2} + 1}{\frac{3x - 1}{x}}$. We want to choose something to multiply numerator and denominator by, so that 2 and x both divide into it. What 'number' should we choose?

$$\dfrac{\frac{x}{2} + 1}{\frac{3x - 1}{x}} = \dfrac{\frac{x}{2} \cdot 2x + 1 \cdot 2x}{\frac{(3x - 1)}{x} \cdot 2x}$$

Multiply numerator and denominator by $2x$.

$$= \dfrac{x^2 + 2x}{2(3x - 1)}$$

The brackets keep the two parts of the numerator together.

C2 Simplify these fractions in the same way.

(a) $\dfrac{1}{a + \frac{1}{2}}$ (b) $\dfrac{x - \frac{1}{3}}{x + \frac{1}{2}}$ (c) $\dfrac{1 - \frac{1}{x}}{\frac{1}{x} - \frac{1}{2}}$

(d) $\dfrac{1 + \frac{x}{y}}{1 + \frac{y}{x}}$ (e) $\dfrac{3 - \frac{a}{b}}{3 - \frac{b}{a}}$ (f) $\dfrac{c + \frac{c}{d}}{c - \frac{c}{d}}$

(g) $\dfrac{\frac{1}{a} - \frac{1}{b}}{\frac{a}{b} - \frac{b}{a}}$ (h) $\dfrac{x - \frac{2}{x} - 1}{x - \frac{6}{x} + 1}$ (i) $\dfrac{\frac{a}{b} + 2 + \frac{b}{a}}{\frac{1}{a} + \frac{1}{b}}$

(j) $\dfrac{\frac{1}{x} + \frac{1}{y} - \frac{1}{xy}}{x + y - 1}$ (k) $\dfrac{\frac{1}{x^2} - \frac{1}{4y^2}}{\frac{1}{2y} + \frac{1}{x}}$

Here is a more interesting example.

Simplify this:

$$\dfrac{2 - \frac{x - 5}{x^2 - x - 2}}{\frac{3}{x + 1}}$$ *Factorise.*

$$= \dfrac{2 - \frac{(x - 5)}{(x - 2)(x + 1)}}{\frac{3}{(x + 1)}}$$ *Use brackets to keep the two parts together.*

Now choose something to multiply numerator and denominator by, so that $(x - 2)(x + 1)$ and $(x + 1)$ both divide into it.

Multiplying by $(x - 2)(x + 1)$, we get

$$\frac{2(x-2)(x+1) - \frac{(x-5)}{(x-2)(x+1)} \times (x-2)(x+1)}{\frac{3}{(x+1)} \times (x-2)(x+1)}$$ Cancelling

$$= \frac{2(x-2)(x+1) - (x-5)}{3(x-2)}$$

$$= \frac{2(x^2 - x - 2) - (x - 5)}{3(x - 2)}$$ Multiplying out, removing brackets, and collecting like terms.

$$= \frac{2x^2 - 2x - 4 - x + 5}{3(x - 2)}$$

$$= \frac{2x^2 - 3x + 1}{3(x - 2)}$$

C3 Simplify the following.

(a) $\dfrac{1 + \frac{1}{x-1}}{1 - \frac{1}{x+1}}$

(b) $\dfrac{b + \frac{bc}{b-c}}{b - \frac{bc}{b+c}}$

(c) $\dfrac{\frac{7}{x+h} - \frac{7}{x}}{h}$

(d) $\dfrac{\frac{1}{6 + a - a^2}}{\frac{1}{a^2 + a - 12}}$

(e) $\dfrac{\frac{16}{x^2} - 1}{\frac{x^2 + 3x - 4}{x}}$

(f) $\dfrac{\frac{x^2 - 3x - 4}{x^2 - 3x}}{\frac{x^2 - 4x}{x + 3}}$

D Adding and subtracting simple fractions

We can only add or subtract fractions if they are the **same type**.

$\frac{3}{7}$ and $\frac{2}{7}$ are the same type. They are both sevenths. $\frac{5}{8}$ and $\frac{1}{8}$ are the same type. They are both eighths.

D1 Copy and complete:

If fractions are the same type then they have . . .

When we are dealing with fractions in algebra, the same rules apply as in arithmetic. So if $\frac{3}{7}$ and $\frac{2}{7}$ are the same type, so are $\frac{3}{7x}$ and $\frac{2}{7x}$.

$\frac{3}{7} + \frac{2}{7} = \frac{5}{7}$ so $\frac{3}{7x} + \frac{2}{7x} = \frac{5}{7x}$

$\frac{3}{7} - \frac{2}{7} = \frac{1}{7}$ so $\frac{3}{7x} - \frac{2}{7x} = \frac{1}{7x}$

D2 The following questions have fractions of the same type. Work them out.

(a) $\frac{2}{9} + \frac{1}{9}$

(b) $\frac{7}{4x} - \frac{2}{4x}$

(c) $\frac{3}{a} + \frac{4}{a}$

(d) $\frac{a}{8} + \frac{3a}{8}$

(e) $\frac{9}{5x} - \frac{4}{5x}$

(f) $\frac{4a}{5} + \frac{2a}{5}$

(g) $\frac{x}{12} + \frac{5x}{12}$

(h) $\frac{7}{a} - \frac{5}{a}$

But, of course, fractions do not normally appear as the same type.

Consider $\frac{1}{4} + \frac{1}{3}$. We have to make these fractions the same type first. Then we can add them.

To make them the same type, look at the denominator. We have to find a positive number that 4 and 3 divide into. The smallest one is 12 – this is called the **lowest common multiple**, or **LCM**.

So $\frac{1}{4} + \frac{1}{3}$ ← (Multiply top and bottom by 4.)

$= \frac{3}{12} + \frac{4}{12}$

$= \frac{7}{12}$

(Multiply top and bottom by 3.)

This can be extended to fractions like these:

$\frac{2a}{3} - \frac{a}{6} + \frac{a}{2}$ ← (Multiply top and bottom by 3.)

(Multiply top and bottom by 2.)

$= \frac{4a}{6} - \frac{a}{6} + \frac{3a}{6}$ (LCM = 6)

$= \frac{6a}{6}$

$= a$

$\frac{1}{2x} - \frac{1}{3x}$ ← (Multiply top and bottom by 2.)

(Multiply top and bottom by 3.)

$= \frac{3}{6x} - \frac{2}{6x}$ (LCM = 6x)

$= \frac{1}{6x}$

D3 What is the LCM of the denominators of these fractions?

(a) $\frac{2a}{5}$ and $\frac{a}{3}$ (b) $\frac{x}{8}$ and $\frac{2x}{3}$ (c) $\frac{2}{y}$ and $\frac{3}{4y}$

(d) $\frac{4}{5x}$ and $\frac{3}{10x}$ (e) $\frac{5a}{2b}$ and $\frac{2a}{b^2}$ (f) $\frac{3}{ab}$ and $\frac{4}{bc}$

(g) $\frac{3}{4x^2}$ and $\frac{4}{3x}$ (h) $\frac{7}{2ab}$ and $\frac{3}{4a^2}$ (i) $\frac{2}{5xy^2}$ and $\frac{3}{2x^2y}$

D4 For each part of question D3, write each as a fraction with the LCM as its denominator.

D5 Simplify the following by taking a common denominator as in the worked examples.

(a) $\frac{5m}{6} - \frac{2m}{3}$ (b) $\frac{5a}{12} + \frac{a}{2}$ (c) $\frac{5}{2a} - \frac{2}{a}$

(d) $\frac{3}{b} + \frac{7}{4b}$ (e) $\frac{2}{3x} + \frac{1}{2x}$ (f) $\frac{4}{5a^2} - \frac{1}{2a^2}$

(g) $\dfrac{1}{2ab} - \dfrac{2}{7bc}$ (h) $\dfrac{x}{3a} + \dfrac{5x}{2a^2}$ (i) $\dfrac{1}{a^2b} - \dfrac{3}{4b}$

(j) $\dfrac{4}{3xy^2} + \dfrac{3}{4x^2y}$ (k) $1 - \dfrac{3x}{4}$ (l) $\dfrac{1}{x} + 3$

(m) $\dfrac{5x}{3} + 1$ (n) $2 + \dfrac{2}{a^2}$ (o) $\dfrac{1}{x^2} - 4$

E Adding and subtracting fractions

When faced with adding or subtracting fractions where the numerator or denominator is not a single term, we use the same method as in section D.

Consider $\dfrac{1}{(a-b)} + \dfrac{1}{(a+b)}$ *Put brackets round both terms to keep them together.*

We need to find a common denominator. That is, we must find a denominator that we can make **both** these denominators into by multiplying.

The common denominator here is $(a-b)(a+b)$ because

$$\dfrac{1}{(a-b)} = \dfrac{(a+b)}{(a-b)(a+b)} \quad \text{and} \quad \dfrac{1}{(a+b)} = \dfrac{(a-b)}{(a-b)(a+b)}$$

We now proceed as before as we now have the same type of fractions.

So $\dfrac{1}{a-b} + \dfrac{1}{a+b} = \dfrac{(a+b)}{(a-b)(a+b)} + \dfrac{(a-b)}{(a-b)(a+b)}$

$= \dfrac{(a+b) + (a-b)}{(a-b)(a+b)} = \dfrac{2a}{(a-b)(a+b)}$ *Collecting like terms.*

E1

Write each pair of fractions with a common denominator.

(a) $\dfrac{1}{x+2}$ and $\dfrac{3}{x+3}$

(b) $\dfrac{3}{x+1}$ and $\dfrac{1}{x+2}$ (c) $\dfrac{4}{x}$ and $\dfrac{3}{x-1}$

(d) $\dfrac{1}{(x-1)(x+1)}$ and $\dfrac{5}{x+1}$ (e) $\dfrac{2}{(x+2)(x-3)}$ and $\dfrac{1}{x+2}$

(f) $\dfrac{7}{x(x+2)}$ and $\dfrac{4}{x}$ (g) $\dfrac{4}{3}$ and $\dfrac{1}{x+1}$

(h) 5 and $\dfrac{3}{x-4}$

(i) $\dfrac{2}{(x-1)(x+1)}$ and $\dfrac{3}{(x+1)(x+2)}$

(j) $\dfrac{1}{(x+3)(x-5)}$ and $\dfrac{3}{x(x+3)}$

E2 Simplify the following by taking a common denominator.

(a) $\dfrac{1}{x+2} + \dfrac{4}{x+3}$

(b) $\dfrac{3}{x+1} - \dfrac{2}{x+2}$

(c) $\dfrac{2}{x+1} + \dfrac{1}{x}$

(d) $\dfrac{4}{2x+1} - \dfrac{2}{x-3}$

(e) $\dfrac{5}{(a+3)(a-3)} + \dfrac{2}{a+3}$

(f) $\dfrac{7}{(2x-3)(2x+3)} - \dfrac{3}{x(2x-3)}$

(g) $\dfrac{9}{(a-1)^2} - \dfrac{3}{(a-1)}$

(h) $\dfrac{1}{(x+4)(x-2)} - \dfrac{1}{(x+4)(x+3)}$

(i) $\dfrac{x+7}{x+2} + \dfrac{25}{(x-3)(x+2)}$

(j) $\dfrac{2x}{(2-x)(2+x)} + \dfrac{1}{x+2} + \dfrac{1}{x-2}$

Simplify $\dfrac{2x+2}{x^2-2x-3} + \dfrac{x-1}{x^2+2x-15}$

(Cartoon: one person says "This is my way to do it", the other says "Here's mine")

Take a common denominator of
$$(x^2 - 2x - 3)(x^2 + 2x - 15)$$

So $\dfrac{2x + 2}{x^2 - 2x - 3} + \dfrac{x - 1}{x^2 + 2x - 15}$

$= \dfrac{(2x + 2)(x^2 + 2x - 15)}{(x^2 - 2x - 3)(x^2 + 2x - 15)} + \dfrac{(x - 1)(x^2 - 2x - 3)}{(x^2 - 2x - 3)(x^2 + 2x - 15)}$

$= \dfrac{(2x + 2)(x^2 + 2x - 15) + (x - 1)(x^2 - 2x - 3)}{(x^2 - 2x - 3)(x^2 + 2x - 15)}$

$= \dfrac{2x^3 + 4x^2 - 30x + 2x^2 + 4x - 30 + x^3 - 2x^2 - 3x - x^2 + 2x + 3}{(x^2 - 2x - 3)(x^2 + 2x - 15)}$

$= \dfrac{3x^3 + 3x^2 - 27x - 27}{(x^2 - 2x - 3)(x^2 + 2x - 15)}$

Factorise $x^2 - 2x - 3 = (x + 1)(x - 3)$
and $x^2 + 2x - 15 = (x - 3)(x + 5)$
and take a common denominator of
$$(x + 1)(x - 3)(x + 5)$$

So $\dfrac{2x + 2}{x^2 - 2x - 3} + \dfrac{x - 1}{x^2 + 2x - 15}$

$= \dfrac{(2x + 2)}{(x + 1)(x - 3)} + \dfrac{(x - 1)}{(x - 3)(x + 5)}$

$= \dfrac{(2x + 2)(x + 5)}{(x + 1)(x - 3)(x + 5)} + \dfrac{(x + 1)(x - 1)}{(x + 1)(x - 3)(x + 5)}$

$= \dfrac{(2x + 2)(x + 5) + (x + 1)(x - 1)}{(x + 1)(x - 3)(x + 5)}$

$= \dfrac{2x^2 + 12x + 10 + x^2 - 1}{(x + 1)(x - 3)(x + 5)}$

$= \dfrac{3x^2 + 12x + 9}{(x + 1)(x - 3)(x + 5)}$

$= \dfrac{3(x^2 + 4x + 3)}{(x + 1)(x - 3)(x + 5)}$

$= \dfrac{3(x + 1)(x + 3)}{(x + 1)(x - 3)(x + 5)}$

$= \dfrac{3(x + 3)}{(x - 3)(x + 5)}$

In fact, the first answer has no mistakes in it, but it is not the **simplest** form of the answer and it cannot be taken further.

> **E3** Since these two answers are really the same, try to convert the second one into the first.
> (**Hint.** Multiply numerator and denominator by some expression.)

The most efficient way to deal with fractions like these is to factorise the denominators first and then take a common denominator.

E4 By first factorising the denominators (if possible), simplify the following.

(a) $\dfrac{1}{x^2 + 2x - 8} - \dfrac{1}{x^2 + 7x + 12}$

(b) $\dfrac{7}{x^2 + x - 12} - \dfrac{1}{x^2 - 5x + 6}$

(c) $\dfrac{1}{a^2 - 2a - 3} - \dfrac{1}{a^2 - 5a - 6}$

(d) $\dfrac{5}{ab + 2b^2} + \dfrac{10}{a^2 + 2ab}$

(e) $\dfrac{1}{x^2 + 9x + 20} + \dfrac{3}{x^2 + 13x + 40}$

(f) $\dfrac{2}{5x - 2} - \dfrac{5x}{25x^2 - 15x + 2}$

(g) $\dfrac{2a - 1}{2a^2 + a - 1} + \dfrac{3a - 4}{3a^2 - 7a + 4}$

(h) $\dfrac{a - 6}{a^2 + a - 42} - \dfrac{a + 3}{a^2 - 4a - 21}$

(i) $\dfrac{2}{x^2 - x} - \dfrac{1}{x^3 - x}$

(j) $\dfrac{1}{x - 4} - \dfrac{2}{x^2 - 6x + 8} + \dfrac{2}{x - 2}$

(k) $\dfrac{9}{x^2 - x - 20} - \dfrac{7}{x^2 + x - 12} - \dfrac{2}{x^2 - 8x + 15}$

(l) $\dfrac{1}{x + 1} - \dfrac{8}{(x^2 - 1)(x^2 + 3)} - \dfrac{2}{x^2 + 3}$

Consolidation 2

A Graphs 2

A1 Ian invests £150 in a savings account which pays 6% interest per annum.

(a) Make a table showing the amount of money in the account at the beginning (0 years), after 1 year, 2 years, and so on up to 6 years.

(b) After how many years would Ian double his money?

(c) Write down a formula for the amount of money in the account after n years.

A2 Sketch the graphs of the following functions for $-2 \leq x \leq 5$.

(a) $y = \frac{100}{x}$ (b) $y = 2^x$ (c) $y = 2(3^x)$

B Indices

Simplify the following.

B1 $(7a^2b)^2$ **B2** $\dfrac{a^3b^5}{4a^4b^3}$

B3 $\dfrac{(3a^4)^2}{(2a)^{-2}}$ **B4** $\dfrac{12a^{\frac{1}{2}}b^{\frac{1}{4}}}{6a^2b}$

Solve the following equations.

B5 $6^x = 216$ **B6** $4^x = 32$ **B7** $3^x = \frac{1}{81}$

B8 Jim has just returned to your class after being ill. He has missed some of the lessons on indices.

How would you explain to him that

(a) $3^0 = 1$ (b) $4^{-3} = \frac{1}{64}$ (c) $16^{\frac{1}{2}} = 4$

C Similar areas and volumes

C1 This tube holds 25 g of toothpaste. A second tube is to be made, which holds 75 g of toothpaste.

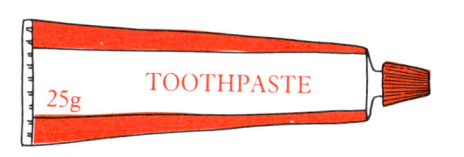

(a) What is the volume factor of the enlargement?

(b) Work out the scale factor of the enlargement, to 2 decimal places.

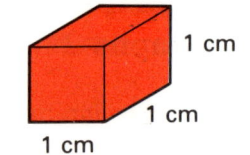

C2 This cube has a volume of 1 cm³. It is to be enlarged to give a cube of length 1 m.

(a) Write down the scale factor of the enlargement.

(b) To cover the small cube, 6 cm² of paper is needed. How much would be needed to cover the large cube?

(c) The large cube weighs 2500 kg. What would you expect the small cube to weigh?

C3 This can is reduced with scale factor 0·9.

(a) Write down how much soup the smaller can will hold.

(b) How much would you expect the smaller can to cost?

D Energy bills

The amount of gas used in a house is measured in units called **therms**. The cost per therm is 15·5p.

In addition to the gas used, a standing charge of £8·25 a quarter (every 3 months) is made.

D1 From 1st November to 31st January, John and Tessie used 2355 therms of gas. How much was this quarter's gas bill?

D2 From 1st May to 31st July, the same couple used only 1020 therms. What was the gas bill this time?

D3 Explain the difference in the amount of gas used in the two quarters.

D4 Alan's last gas bill amounted to £253·92. How many therms of gas had he used?

D5 Ann pays her gas bill monthly. If she used 137 therms last month, calculate her bill.

D6 Last May, Ann paid a bill of £37·31. How many therms did she use?

*Remember the standing charge for **one** month!*

E Circles

E1 A tangent kite is constructed in a circle of radius 6 cm, from a point, B, 10 cm from the centre, O.

Calculate the area of the kite.

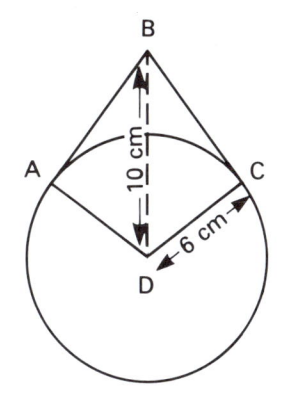

E2 Calculate the sizes of the marked angles.

(a)

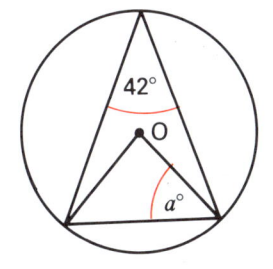

(b)

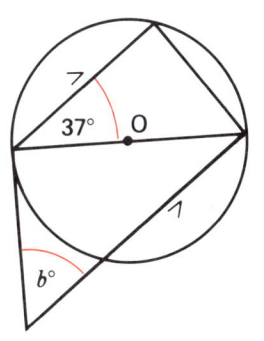

E3 Calculate the radius of this circle.

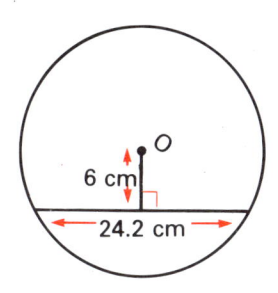

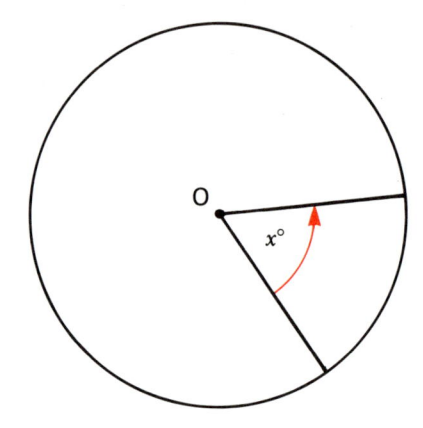

E4 The radius of the larger circle in this diagram is three times longer than the radius of the smaller circle.

What angle $x°$ would give a sector of the large circle which is equal in area to the small circle?

F Algebraic fractions

Simplify these.

F1 $\dfrac{7abc}{21a^2b}$ **F2** $\dfrac{x-a}{a-x}$

F3 $\dfrac{x^2+5x+6}{2x^2+5x+2}$ **F4** $\dfrac{x}{y}-\dfrac{y}{x}$

F5 $\dfrac{3}{a}+\dfrac{1}{a+4}$ **F6** $\dfrac{y}{2x}+\dfrac{4}{x^2}$

F7 $\dfrac{\frac{3a}{b^2}}{\frac{b}{6a}}$ **F8** $\dfrac{\frac{x}{a+b}}{\frac{b}{x+y}}$

F9 $\dfrac{2+\frac{2}{x}}{1+\frac{1}{x}}$ **F10** $\dfrac{\frac{1}{x}+\frac{1}{y^2}}{x+\frac{1}{y}}$

9 The sine rule and the area of a triangle

A The problem

This is a roof section from a house.

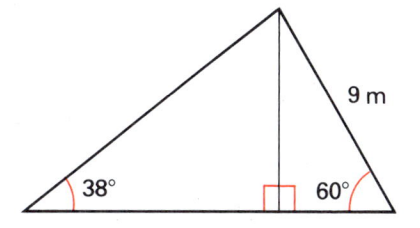

The side inclined at 60° to the horizontal is 9 m long. How long is the side inclined at 38°?

Using triangle BDC we can work out DC.

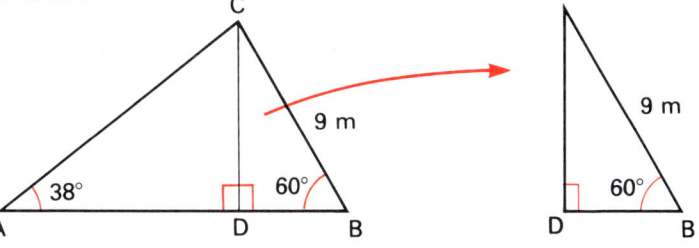

A1 Copy and complete:

sin 60° = ...
0·866 = ...
DC = ...

So the diagram now looks like this.

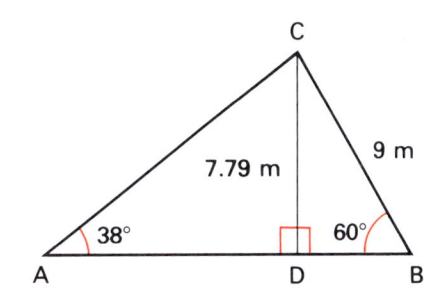

A2 Now use triangle ACD to work out the length of AC.

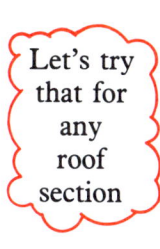

Let's try that for any roof section

| The diagram | 1 | Label the sides |

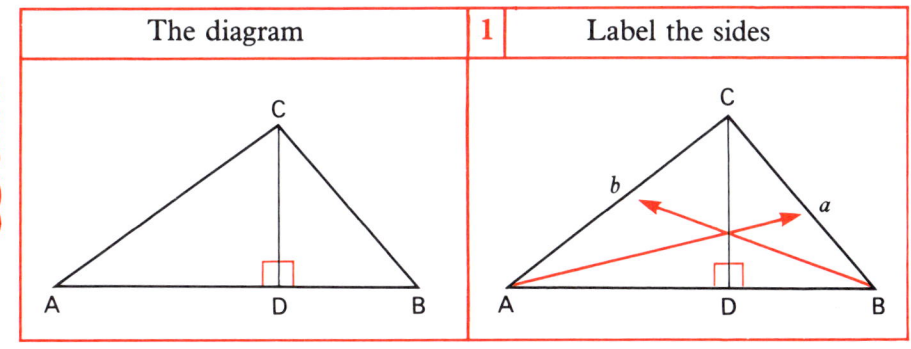

99

| 2 | Use triangle BCD to find DC |

B C D *a*

$\sin B = \dfrac{CD}{a}$

$CD = a \sin B$

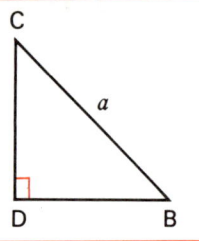

| 3 | Use triangle ACD to find AC |

$\sin A = \dfrac{a \sin B}{b}$

$b \sin A = a \sin B$

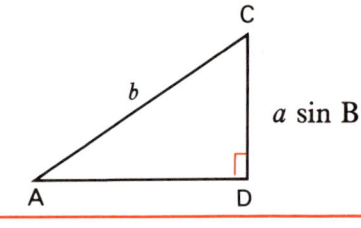

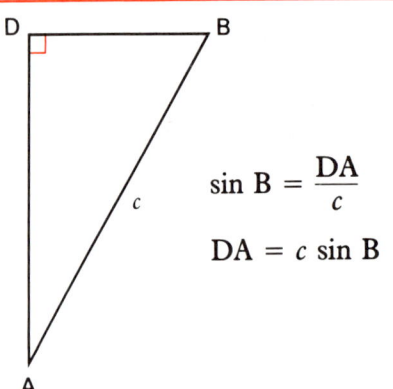

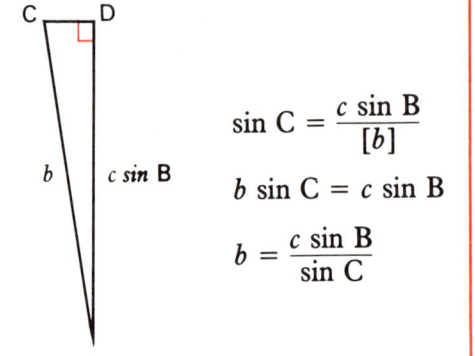

$b = \dfrac{a \sin B}{\sin A}$

| 4 | Re-arrange the result |

$b = \dfrac{a \sin B}{\sin A}$ $\dfrac{b}{\sin B} = \dfrac{a}{\sin A}$ *Dividing both sides by sin B*

Now do the same to find AD and then AC.

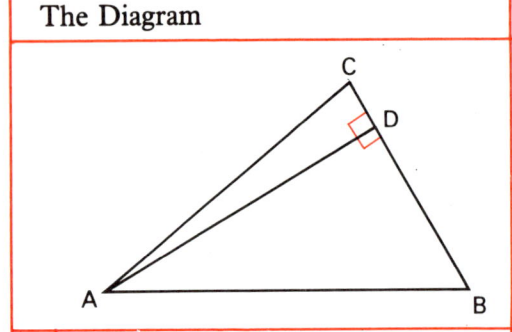

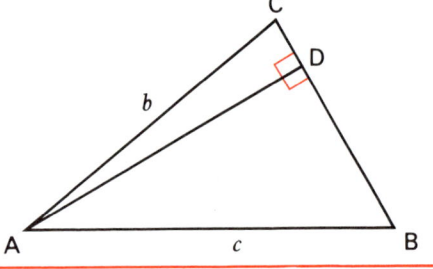

| The Diagram | | 1 | Label the sides |

| 2 | Use triangle BAD to find DA |

$\sin B = \dfrac{DA}{c}$

$DA = c \sin B$

| 3 | Use triangle ACD to find AC |

$\sin C = \dfrac{c \sin B}{[b]}$

$b \sin C = c \sin B$

$b = \dfrac{c \sin B}{\sin C}$

4	Rearrange the result
	$b = \dfrac{c \sin B}{\sin C}$ $\dfrac{b}{\sin B} = \dfrac{c}{\sin C}$ ← Dividing both sides by sin B

Combining the results we have

$$\dfrac{a}{\sin A} = \dfrac{b}{\sin B} = \dfrac{c}{\sin C}$$

This is called the **sine rule**.

Notice the pattern of letters in the sine rule.
Each fraction is like this:

$$\dfrac{\text{Side of triangle}}{\text{Sine of angle opposite this side}}$$

A3 Write down the sine rule for the following triangles. The first one has been started for you.

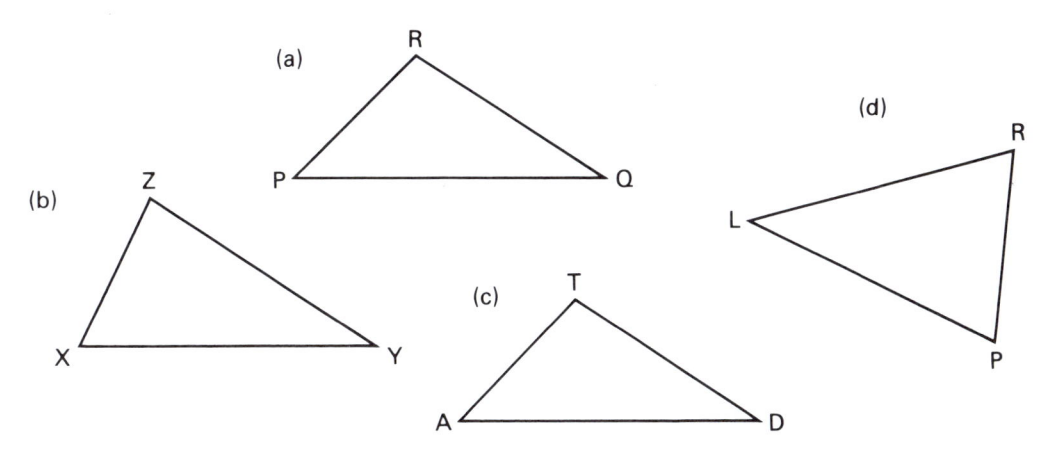

(a) $\dfrac{p}{\sin P} = \dfrac{q}{\ldots} = \dfrac{\ldots}{\sin R}$

A4 Would you use the sine rule in a right-angled triangle? Explain your answer.

Using the sine rule

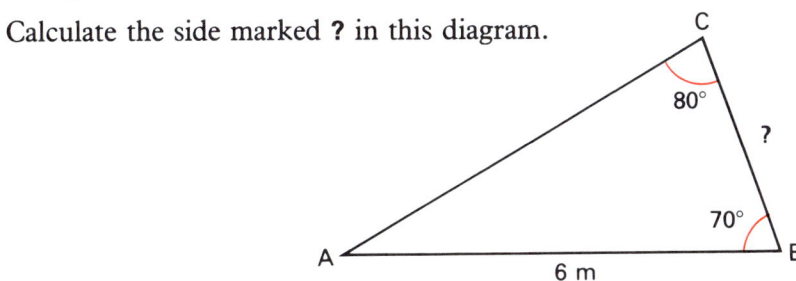

Calculate the side marked **?** in this diagram.

1	Calculate the third angle of the triangle if possible.	2	Write down the sine rule and tick the known measures.

Since we have two angles of 80° and 70°, the angle at A is 30°.

For triangle ABC, the sine rule is

$$\frac{a?}{\sin A} = \frac{b}{\sin B} = \frac{c}{\sin C}$$

We will use

$$\frac{a}{\sin A} = \frac{c}{\sin C}$$

because we know 3 of the 4 measures

3	Substitute values and solve equation

$$\frac{a?}{\sin A} = \frac{c}{\sin C}$$

Cross-multiply

$$\frac{a}{\sin 30°} = \frac{6}{\sin 80°}$$

$$a \sin 80° = 6 \sin 30°$$

$$a = \frac{6 \sin 30°}{\sin 80°}$$

$$a = \frac{6 \times 0{\cdot}5}{0{\cdot}985}$$

$$a = 3{\cdot}05 \text{ m}$$

So BC = 3·05 m

A5 Now try these. Set out your working like the example.

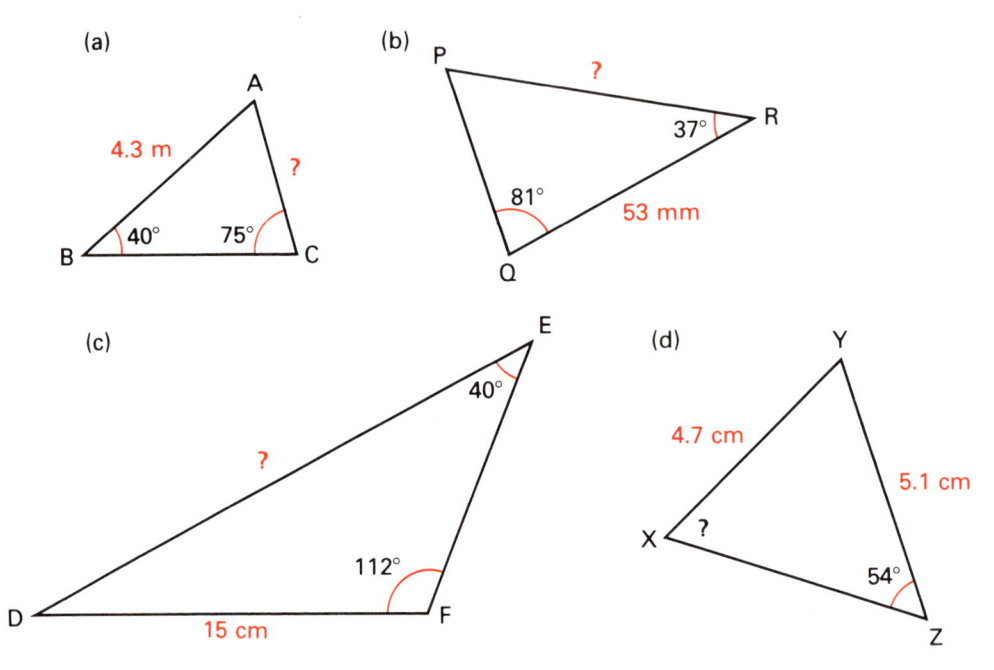

(a) Triangle with B (40°), C (75°), A; side BA = 4.3 m, side AC = ?

(b) Triangle PQR with angle at Q = 81°, angle at R = 37°, side QR = 53 mm, side PR = ?

(c) Triangle DEF with angle at E = 40°, angle at F = 112°, side DF = 15 cm, side DE = ?

(d) Triangle XYZ with angle at Z = 54°, side YX = 4.7 cm, side YZ = 5.1 cm, angle at X = ?

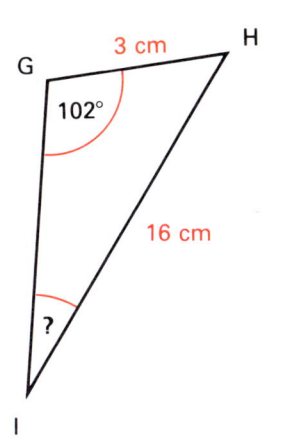

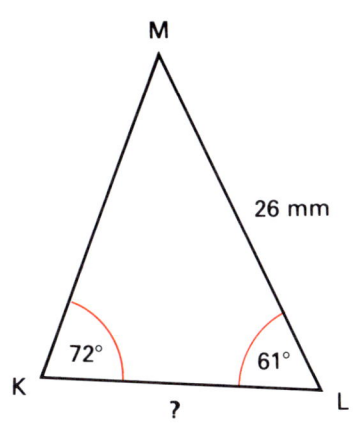

A6 (a) For the triangle below there are two possible values for angle B. Find both of them.

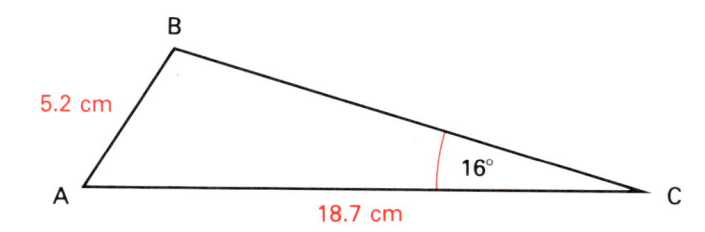

(b) If triangle ABC is given as below, show that there is only one possible value for angle B and find that value.

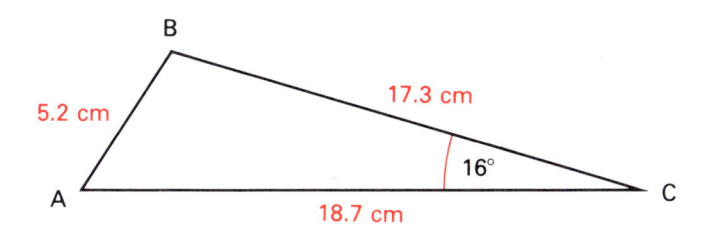

A7 I. Skelpem, the famous international golfer, got a hole in two. His first shot was at an angle of 8° to the direct line to the hole.

If the hole was 200 m long, and the angle between his first shot and his second shot was 102°, calculate how far he was away from the hole after his first shot.

A8 Mrs Larchlap's 1·2 m high fence has been almost blown down by the wind. She wants to tie a piece of strong cord to it and to a peg 1·2 m from the base of the fence.

The fence has been blown so that it has moved 20° away from the vertical, as shown in the sketch.

What is the minimum length of cord that Mrs Larchlap should buy if it can be bought in units of 20 cm and she allows 40 cm for tying a knot?

A9 Jess has a wild dog, Barker, on her farm. She is going to make a pen for it against a wall which bends at an angle of 160°. She has only 100 m of fencing, and wants to start the fence at the gate, 50 m away from the bend in the wall.

How far past the bend will the fence reach?

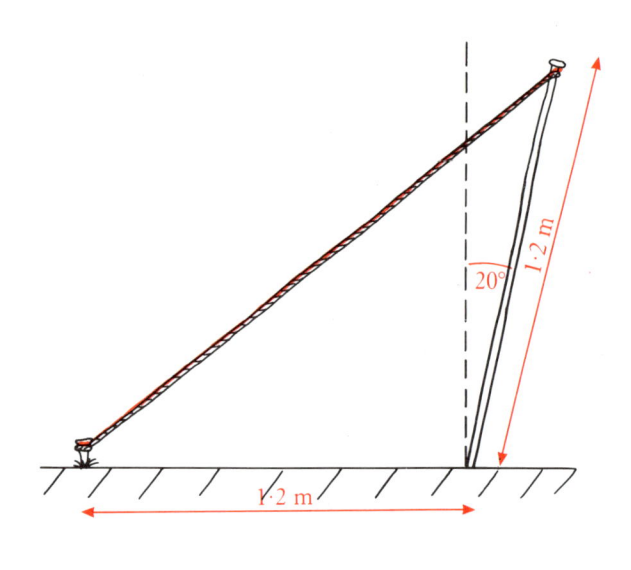

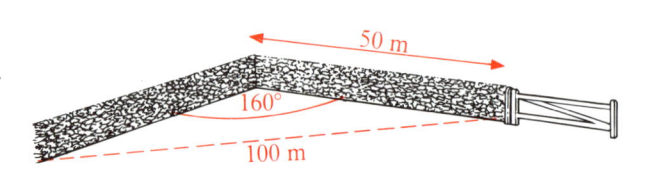

A10 (a) Archie and Pascale are standing on opposite sides of the Eiffel Tower.
The distance between them is 230 m and the angles of elevation of the top of the tower are 72° and 66°. If they are standing on level ground, calculate the height of the Eiffel Tower.

(b) If Archie and Pascale were both standing on the same side of the Tower, in line with it, calculate the distance between them.

A11 Barbara is an orienteer. She runs for 150 m on a bearing of 070° then changes course and runs for another 170 m. Her final position is on a bearing of 056° from her starting point. Assuming Barbara always runs in a straight line, calculate how far she finishes from her starting point.

A12 From a point on Princes Street the angle of elevation of the top of the flagpole on Edinburgh Castle is 45° and the angle of elevation of the bottom of the flagpole is 42°. If the flagpole is 10 m long, calculate the height of the top of the flagpole above the level of Princes Street.

B The area of a triangle

Horatio Smith, the famous sailor, is designing sails for his new yacht.

The mainsail is drawn below.

Horatio needs to work out the area of the sail so he can order material to make it up.
Horatio knows that the area of a triangle can be worked out using the formula

   Area = $\frac{1}{2}$ × base × height
or   A = $\frac{1}{2}bh$

so all he needs to find is the perpendicular height of the sail.

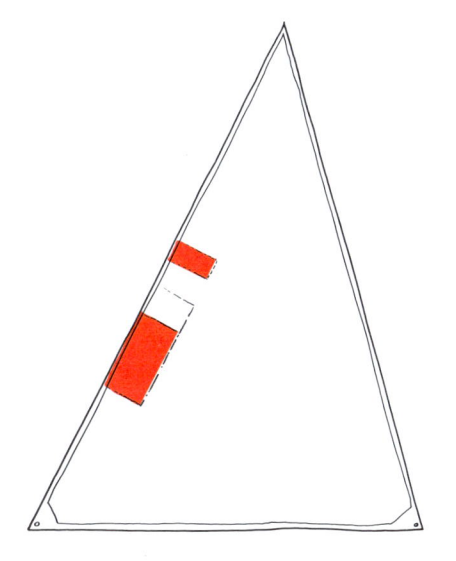

> Let's try that for any triangle FDE

B1 Copy and complete the following steps.

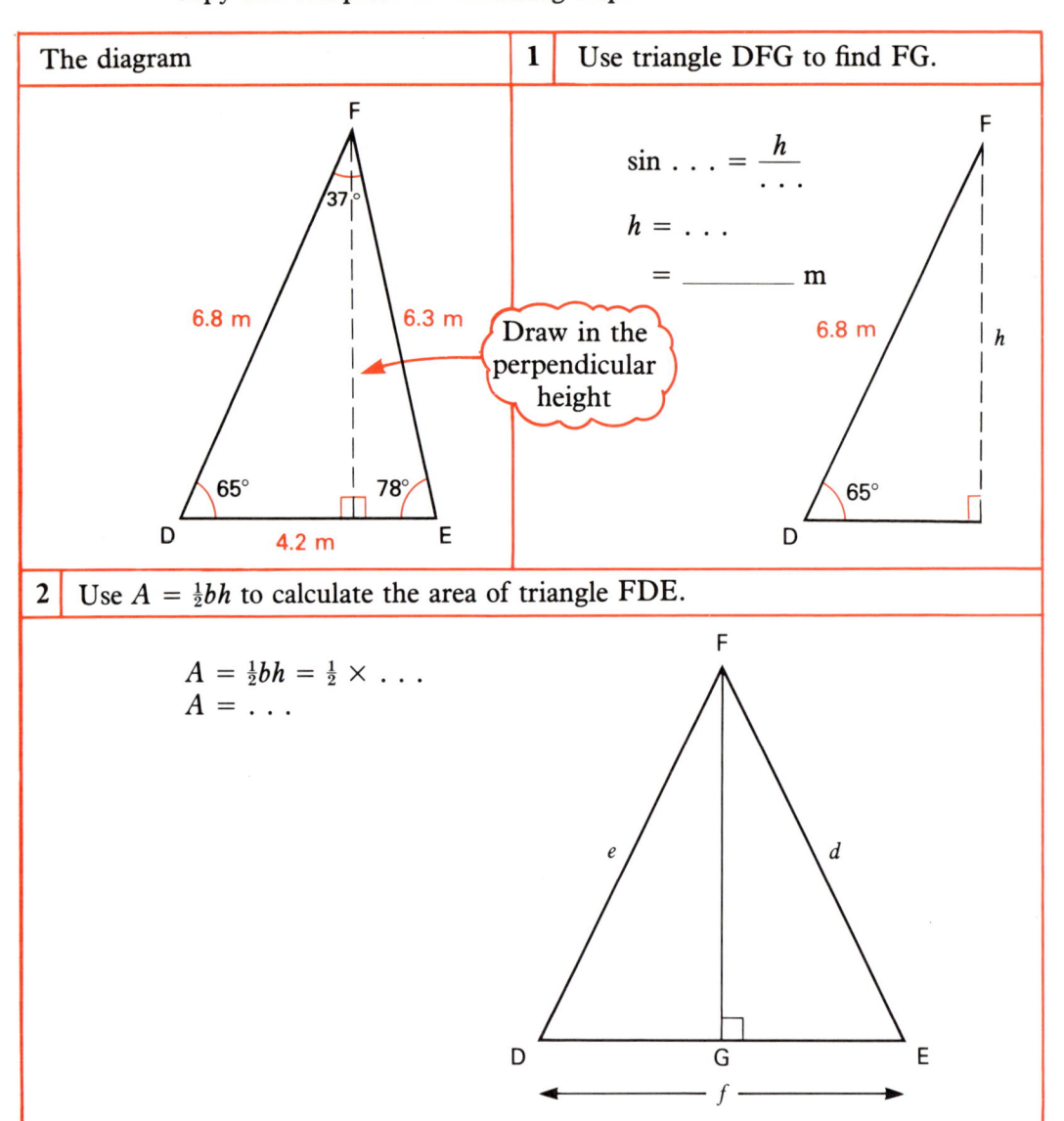

The diagram	1 Use triangle DFG to find FG.
Triangle FDE with F at top, angle at F = 37°, DF = 6.8 m, FE = 6.3 m, angle D = 65°, angle at E (with perpendicular from F to DE meeting at right angle) = 78°, DE = 4.2 m. "Draw in the perpendicular height"	$\sin \ldots = \dfrac{h}{\ldots}$ $h = \ldots$ $ = \underline{}$ m Right triangle with F at top, DF = 6.8 m, angle D = 65°, height h.

2 Use $A = \tfrac{1}{2}bh$ to calculate the area of triangle FDE.

$A = \tfrac{1}{2}bh = \tfrac{1}{2} \times \ldots$
$A = \ldots$

Triangle FDE with perpendicular FG, sides labelled e (DF), d (FE), base f (DE).

B2 Draw triangle DFG. Copy and complete the following.

In triangle DFG
$\sin D = \dfrac{FG}{\ldots}$
So FG = ...

In triangle DEF
$A = \tfrac{1}{2}bh$
$ = \tfrac{1}{2} \times f \times \ldots$
So $A = \tfrac{1}{2} \times \ldots$

This is a formula for the area of any triangle DEF.

B3 Note that we need to know three things about the triangle – what are they?

B4 Take the same triangle DEF as in question B2 and draw in the perpendicular height from D onto FE.

Using the same technique as before, try to show that

Area = $\frac{1}{2}de \sin F$ and   Area = $\frac{1}{2}df \sin E$

C Using the formula

Look at the formulas we have now worked out for the area of triangle DEF.

$$\frac{1}{2}de \sin F \quad \frac{1}{2}df \sin E \quad \frac{1}{2}ef \sin D$$

C1 What pattern do you notice in them?

C2 Use this pattern to write down **three** formulas for the area of each of these triangles:

(a) Triangle ABC   (b) Triangle PQR   (c) Triangle XYZ

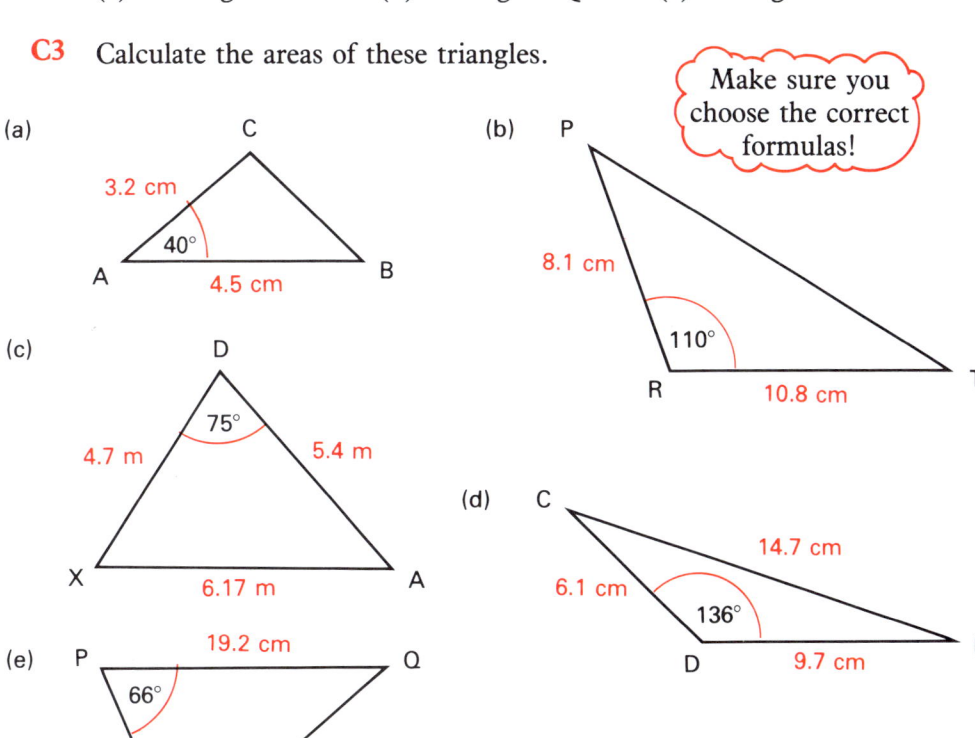

C3 Calculate the areas of these triangles.

Make sure you choose the correct formulas!

(a) Triangle with C at top, A bottom-left, B bottom-right; AC = 3.2 cm, angle A = 40°, AB = 4.5 cm.

(b) Triangle with P at top, R bottom-left, T bottom-right; PR = 8.1 cm, angle R = 110°, RT = 10.8 cm.

(c) Triangle with D at top, X bottom-left, A bottom-right; DX = 4.7 m, angle D = 75°, DA = 5.4 m, XA = 6.17 m.

(d) Triangle with C at top-left, D at bottom, M at right; CD = 6.1 cm, angle D = 136°, CM = 14.7 cm, DM = 9.7 cm.

(e) Triangle with P at top-left, Q at top-right, D at bottom; PQ = 19.2 cm, angle P = 66°, angle at lower vertex = 74°, QD = 18.2 cm.

107

C4 Calculate the areas of these triangles.

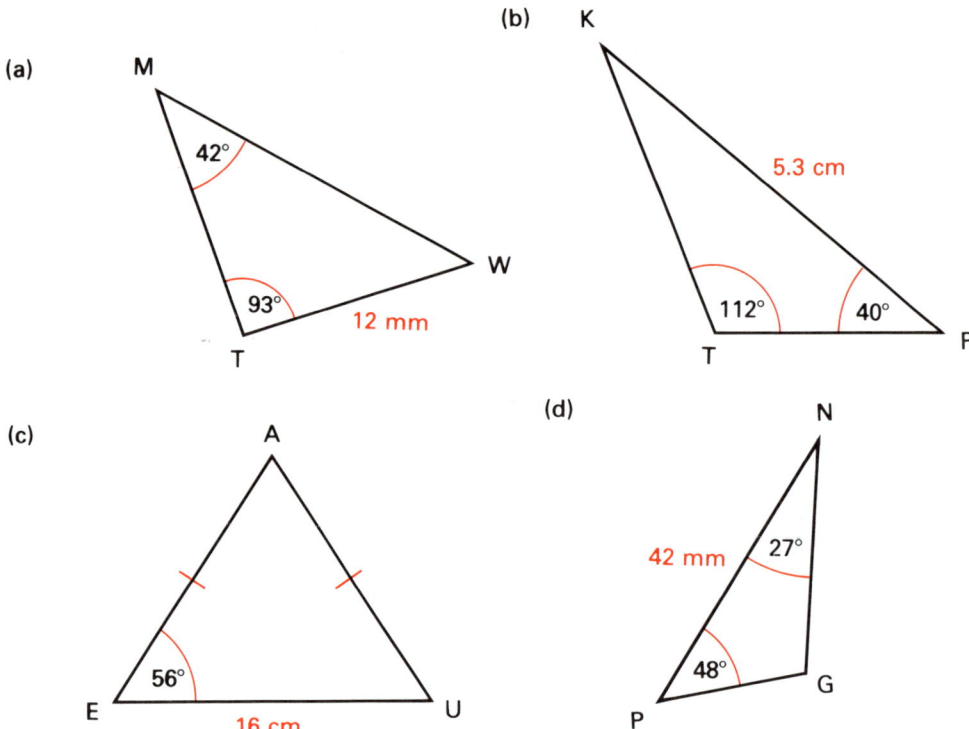

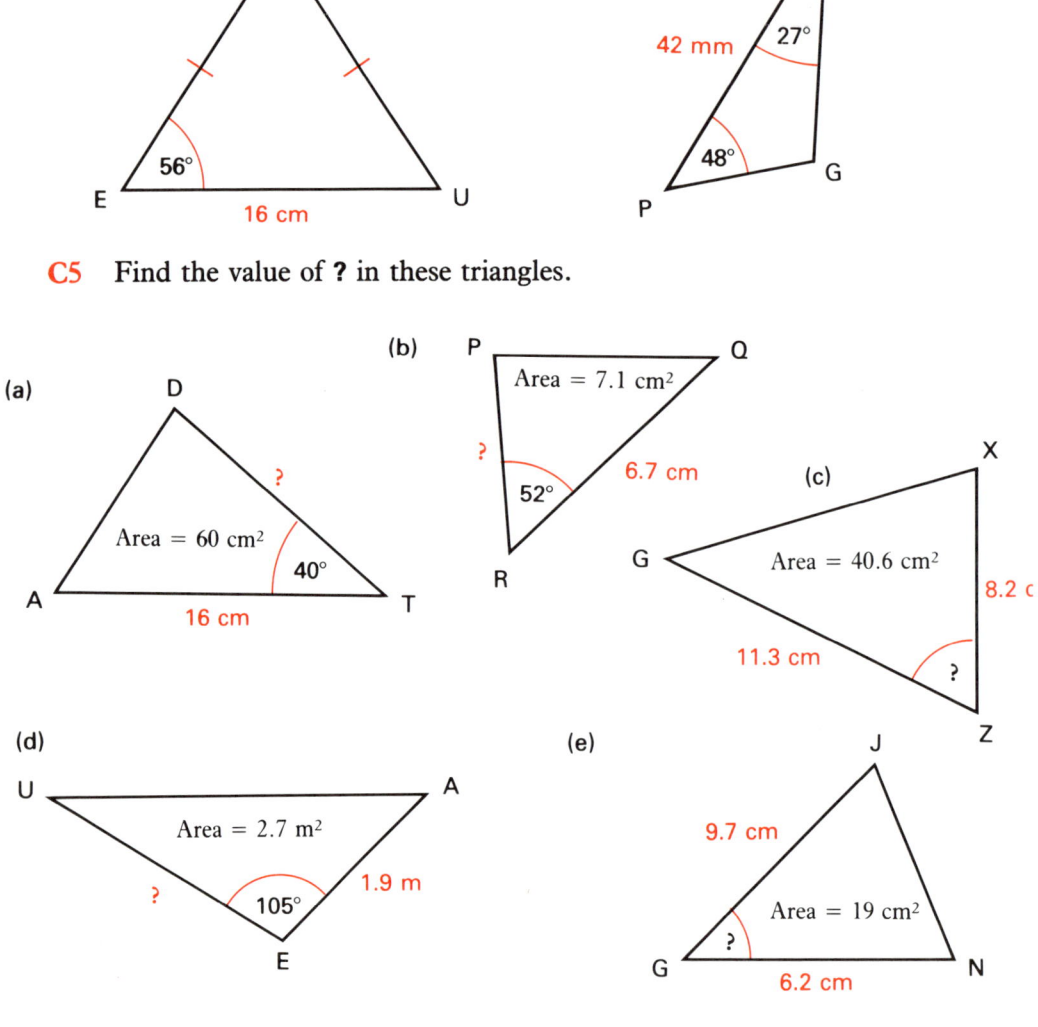

(a) Triangle MTW: angle M = 42°, angle T = 93°, TW = 12 mm

(b) Triangle KTP: KP = 5.3 cm, angle T = 112°, angle P = 40°

(c) Triangle EAU: AE = AU (isosceles), angle E = 56°, EU = 16 cm

(d) Triangle PNG: PN = 42 mm, angle N = 27°, angle P = 48°

C5 Find the value of **?** in these triangles.

(a) Triangle ADT: Area = 60 cm², AT = 16 cm, angle at T = 40°, AD = ?

(b) Triangle PQR: Area = 7.1 cm², PR = 6.7 cm, angle R = 52°, PQ = ?

(c) Triangle GXZ: Area = 40.6 cm², GX = 11.3 cm, XZ = 8.2 cm, angle at Z = ?

(d) Triangle UAE: Area = 2.7 m², AE = 1.9 m, angle E = 105°, UE = ?

(e) Triangle GJN: Area = 19 cm², GJ = 9.7 cm, GN = 6.2 cm, angle G = ?

108

C6 Look back at question A9. Calculate the area of Barker's pen.

C7 Megan likes flying unusual kites. She wants to make a new kite which is made from an isosceles triangle and a scalene triangle. This diagram shows its dimensions.

A scalene triangle's sides are all of different lengths.

Calculate how much material Megan needs to make her new kite.

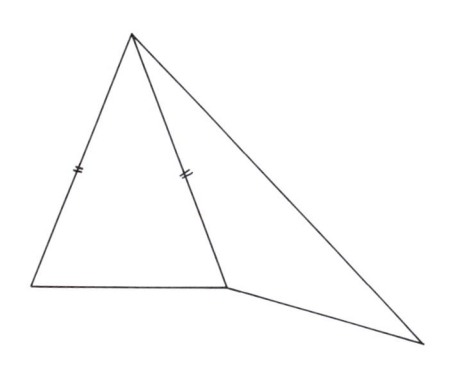

C8 A painter is estimating how much paint he will need to paint the wooden gable end of a modern house.

He measures the sides of the triangular part to be 8·2 m, 4·3 m and 6·7 m.

The paint he is going to use covers between 8 m² and 10 m² per litre. Calculate the maximum and minimum amount of paint he will need (to the nearest whole number of litres) to give the gable two coats of paint.

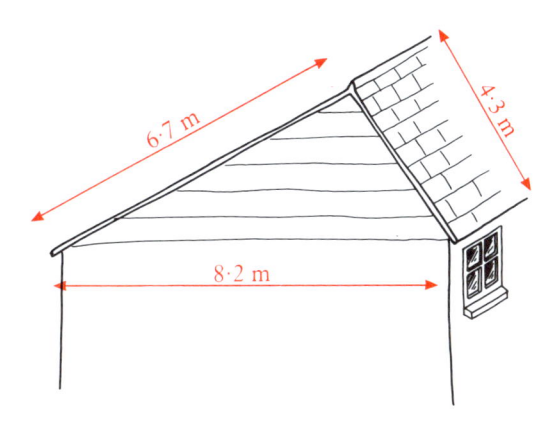

10 Quadratic equations

A L-shapes

Bill designs signs. He is trying to design a letter L so that it fits this description.

The letter is to be made from 36 cm² of fluorescent material. Bill has to calculate the width of his letter L.

This height must be 6 cm

This length must be 3 cm

We are trying to find this side. We will call it x cm.

A1 Copy this sketch of the letter. Work out a formula for its area by splitting it up into rectangles.

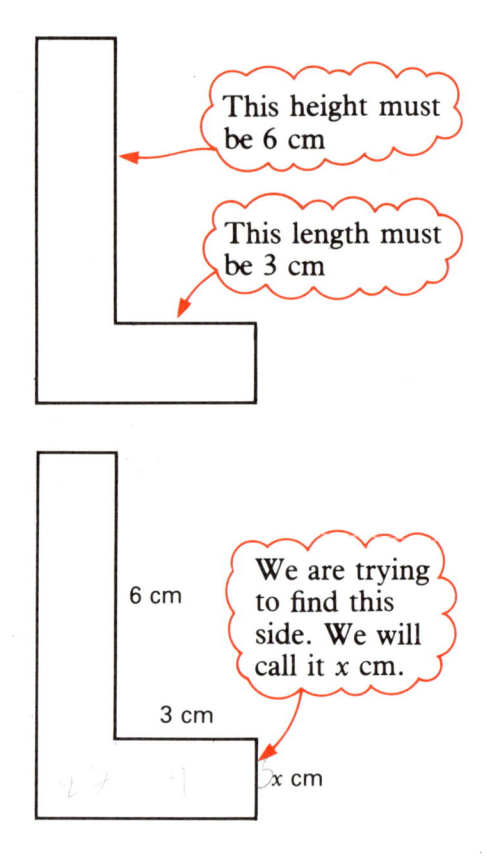

A2 Now write the formula for the area in a factorised form.

We know that the area of the letter must be 36 cm², so we can form an equation in x:

$$x(x + 9) = 36$$

A3 There are two solutions to this equation. Find them and explain why one of them is not valid.

Bill changes his mind about the proportions of the L. This time he wants to design it like this.

110

A4 (a) Find the formula for the area of this L-shape and write it in factorised form.

(b) The area is 36 cm² again. Find the two solutions and explain why one of them is not valid.

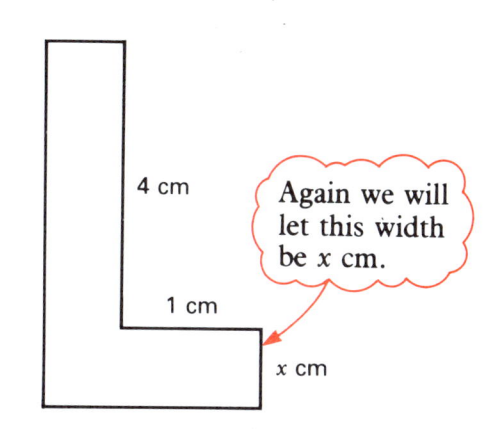

Bill decides to try another design. It still has an area of 36 cm².

A5 Using the same method as before, find a formula for the area of the L and form an equation. Can you find the values of x this time?

Equations like

$x(x + 7) = 36$

are called **quadratic equations** because when the brackets are multiplied out you get a quadratic form.

$$x(x + 7) = 36$$
$$x^2 + 7x = 36$$
or $$x^2 + 7x - 36 = 0$$

111

B Solving quadratic equations by completing the square

Consider the simple quadratic equation
$$x^2 = 9$$
The solutions are $x = 3$ and $x = -3$.
We can write this as $x = \pm 3$.

Now take the quadratic equation
$$(x + 2)^2 = 9$$

*This is called a **perfect square**.*

If we compare this to the one before we have

(something)$^2 = 9$ $(x + 2)^2 = 9$
something $= 3$ or something $= -3$ $x + 2 = 3$ or $x + 2 = -3$
and so $x = 1$ or $x = -5$

B1 Use this method to solve the following quadratic equations. In parts (g) and (h), work to 4 s.f. and round to 3 s.f. at the end.

(a) $(x + 7)^2 = 16$ (b) $(x - 2)^2 = 25$
(c) $(x - 4)^2 = 100$ (d) $(2x - 1)^2 = 81$
(e) $(5 - x)^2 = 64$ (f) $(x - 5)^2 = 4$
(g) $(x + 3)^2 = 7$ (h) $(1 - 3x)^2 = 19$

If we have a quadratic equation and the left side is a perfect square, we can solve it using this method.

But what about equations like $x^2 + 6x = 40$?

First of all we must investigate the pattern involved when we square out brackets.

B2 Multiply out the brackets in the following.

(a) $(x + 1)^2$ (b) $(x - 2)^2$ (c) $(x + 3)^2$
(d) $(x - 4)^2$ (e) $(x - 5)^2$ (f) $(x + 9)^2$
(g) $(x - 8)^2$ (h) $(x + 6)^2$ (i) $(x - 7)^2$

B3 Explain in words the pattern in your answers to question B2.

B4 Using the pattern, copy and complete this table.

	Term required to make a perfect square	The expression is then	The expression as a perfect square
(a) $x^2 + 4x + \ldots$			
(b) $x^2 - 12x + \ldots$			
(c) $x^2 - 18x + \ldots$			
(d) $x^2 + \ldots + 64$			
(e) $x^2 - \ldots + 100$			
(f) $x^2 - 7x + \ldots$			
(g) $x^2 - \ldots + \frac{9}{4}$			
(h) $x^2 + 11x + \ldots$			

Let us now go back to Bill's 'perfect L' and consider the quadratic equation he had to solve:

$x^2 + 7x = 36$

We can change the left side into a perfect square by using the method of question B4.

Adding $\frac{49}{4}$ to both sides, we get

$x^2 + 7x + \frac{49}{4} = 36 + \frac{49}{4}$
$(x + \frac{7}{2})^2 = \frac{144 + 49}{4}$
$(x + \frac{7}{2})^2 = \frac{293}{4}$

We can solve this like question B1.

$x + \frac{7}{2} = 8 \cdot 559$ or $x + \frac{7}{2} = -8 \cdot 559$
$x + 3 \cdot 5 = 8 \cdot 559$ or $x + 3 \cdot 5 = -8 \cdot 559$
$x = 8 \cdot 559 - 3 \cdot 5$ or $x = -8 \cdot 559 - 3 \cdot 5$
$x = 5 \cdot 059$ or $x = -12 \cdot 059$
$x = 5 \cdot 06$ or $x = -12 \cdot 1$ (rounded to 3 s.f.)

Since $\sqrt{\frac{293}{4}}$ does not work out exactly we work with 4 s.f. and round to 3 s.f. at the end.

This would be the width of the L-shape and would be taken to be 5·1 cm or 5·06 cm, depending on the degree of accuracy required by the designer.

This answer is not valid as the width of the L-shape could not be negative.

B5 Use this method (called **completing the square**) to solve these quadratic equations. If necessary work to 4 s.f. and round to 3 s.f. at the end.

(a) $x^2 + 6x = 7$
(b) $x^2 - 10x = 24$
(c) $x^2 - 8x = -7$
(d) $x^2 - 12x + 11 = 0$
(e) $x^2 + 9x = 4$
(f) $x^2 = 6x - 4$
(g) $3 = x^2 + 5x$
(h) $5 = x^2 + 3x$

B6 Try to solve $x^2 + 4x = -5$ using the method of completing the square. Why does it not work?

C Solving quadratic equations using the formula

The technique that is now available to solve quadratic equations has two flaws:
- Sometimes we have to find the square root of a negative value. Since we cannot find this, there is no valid solution (as in question B6).
- The coefficient of the x^2 term is always 1. This can be overcome as follows:

$2x^2 + 12x = 8$ — Divide everything by 2. — We now proceed as before.
$x^2 + 6x = 4$
$x^2 + 6x + 9 = 4 + 9$
$(x + 3)^2 = 13$
$x + 3 = 3·606$ or $x + 3 = -3·606$
$x = 0·606$ (to 3 s.f.) or $x = -6·606 = -6·61$ (to 3 s.f.)

C1 Solve these quadratic equations, rounding your answers to 1 decimal place where necessary. (This means that you will need to work to 2 d.p. and round to 1 d.p. at the end.)

(a) $2x^2 + 8x = 10$
(b) $3x^2 - 30x = -27$
(c) $2x^2 - 6x = -2$
(d) $4x^2 + 12x = 8$
(e) $5x^2 + 10x = 10$
(f) $2x^2 = 20x + 4$
(g) $6x = 3x^2 - 6$
(h) $4x^2 - 6x = -1$
(i) $3x^2 - 12x = 7$
(j) $2 = 3x^2 + 5x$

We can now investigate our method of completing the square to solve quadratic equations and apply it to *any* quadratic equation.

In *Mathematics for Credit 1* the general form for quadratic expressions was given as

$ax^2 + bx + c$

where a, b and c are real numbers and $a \neq 0$.

Using the same method as in question C1, we can solve the general quadratic equation

$$ax^2 + bx + c = 0$$
$$ax^2 + bx = -c$$

Subtract c from both sides.

$$x^2 + \frac{b}{a}x = -\frac{c}{a}$$

Divide everything by a. There is no problem since $a \neq 0$.

$$x^2 + \frac{b}{a}x + \left(\frac{b}{2a}\right)^2 = \left(\frac{b}{2a}\right)^2 - \frac{c}{a}$$

$$(x + \frac{b}{2a})^2 = \left(\frac{b}{2a}\right)^2 - \frac{c}{a}$$

We now complete the square as before.

$$= \frac{b^2}{4a^2} - \frac{4ac}{4a^2}$$

$$= \frac{b^2 - 4ac}{4a^2}$$

Taking the square root

$$x + \frac{b}{2a} = +\sqrt{\frac{b^2 - 4ac}{4a^2}} \text{ or } x + \frac{b}{2a} = -\sqrt{\frac{b^2 - 4ac}{4a^2}}$$

which can be written as

$$x + \frac{b}{2a} = \pm\sqrt{\frac{b^2 - 4ac}{4a^2}}$$

$$= \frac{\pm\sqrt{b^2 - 4ac}}{\sqrt{4a^2}}$$

$$= \pm\frac{\sqrt{b^2 - 4ac}}{2a}$$

$$x = -\frac{b}{2a} \pm \frac{\sqrt{b^2 - 4ac}}{2a}$$

$$x = \frac{-b \pm \sqrt{b^2 - 4ac}}{2a}$$

This is a formula for finding the solutions to the general quadratic equation.

A **mnemonic** (ni-mon-ik) is a verse of rhyme to help remember a rule. There is an eighteenth-century mnemonic for this formula. It goes like this;

> From square of b
> Take $4ac$,
> Square root extract
> And b subtract,
> Divide by $2a$
> You'll get x alway.

Example

Solve $5x^2 + 8x - 2 = 0$ using the quadratic formula. Give your answers to 1 decimal place.

Step 1 Put the equation in standard form and identify a, b and c.

$5x^2 + 8x - 2 = 0$ is in the same form as $ax^2 + bx + c = 0$ so it is in standard form.

By comparison $a = 5$, $b = 8$, $c = -2$

Step 2 Write down the formula and substitute values

$$x = \frac{-b \pm \sqrt{b^2 - 4ac}}{2a} = \frac{-8 \pm \sqrt{8^2 - 4 \times 5 \times (-2)}}{10}$$

$$= \frac{-8 \pm \sqrt{64 + 40}}{10}$$

$$= \frac{-8 \pm \sqrt{104}}{10}$$

Step 3 Split the working into two parts (one for $+\sqrt{104}$, the other for $-\sqrt{104}$) and work to an accuracy one figure more than finally required.

$(+)$

$$x = \frac{-8 + \sqrt{104}}{10}$$

$$= \frac{-8 + 10 \cdot 20}{10}$$

$$= \frac{2 \cdot 20}{10}$$

$$= 0 \cdot 220$$

or

$(-)$

$$x = \frac{-8 - \sqrt{104}}{10}$$

$$= \frac{-8 - 10 \cdot 20}{10}$$

$$= \frac{-18 \cdot 20}{10}$$

$$= -1 \cdot 82$$

Step 4 Round the answers to the required degree of accuracy.

$x = 0 \cdot 2$ or $-1 \cdot 8$

C2 Check that the formula works by using it to solve questions B5(a) and C1(a) again and comparing with the answers you found before.

C3 Solve the following quadratic equations where possible using the quadratic formula. Round your answers to 2 decimal places. If any equation cannot be solved give the reason why.

(a) $2x^2 + 4x + 1 = 0$ (b) $3x^2 + 2x - 4 = 0$

(c) $x^2 + x + 1 = 0$ (d) $2x^2 = 5x + 7$

(e) $9x = x^2 - 2$ (f) $x(x + 5) = 2$

(g) $8 - x(x + 4) = 0$ (h) $3(x^2 + 7) = 4x$

(i) $x(x - 2) = 3(1 - x)$ (j) $4x(1 - 3x) = 3(x - 4)$

D Solving quadratic equations by factorisation

D1 If you multiply any number by zero, what answer do you always get?

D2 If $4a = 0$, what value must a have? Give a reason for your answer.

Remember $4a$ means $4 \times a$.

D3 If $4(x - 7) = 0$, what value must x have? Give a reason for your answer.

D4 If $(y + 2)5 = 0$, what value must y have? Give a reason for your answer.

D5 (a) If $(x + 2)(x - 7) = 0$ and **x is positive**, what value must x have? Check that this value satisfies the equation.

(b) If $(x + 2)(x - 7) = 0$ and **x is negative**, what value must x have? Check that this value satisfies the equation.

D6 If $(x + 5)(x + 9) = 0$, what are the two possible values for x? Check each solution satisfies the equation.

The left sides in questions D5 and D6 are both quadratic expressions when the brackets are multiplied out. This leads to another way of solving quadratic equations, by factorisation.

Consider $x^2 + 2x - 15 = 0$.
Factorise $x^2 + 2x - 15$

	−15	15	5	−5
	1	−1	−3	3
Adding	−14	14	2	−2

So $x^2 + 2x - 15 = (x + 5)(x - 3)$

Our equation becomes $(x + 5)(x - 3) = 0$
So either $x + 5 = 0$ or $x - 3 = 0$
            $x = -5$ or $x = 3$
The two solutions are -5 and 3.

It is always a good idea to check each solution to make sure it satisfies the equation.

$x = 3$: $x^2 + 2x - 15$
           $= 9 + 6 - 15$
           $= 0$

This is true so $x = 3$ satisfies the equation.

D7 Check the other solution.

D8 Solve the following quadratic equations by factorisation. Check at least one solution in each case.

(a) $x^2 + 2x - 35 = 0$ (b) $x^2 + 7x + 12 = 0$

(c) $0 = x^2 - 6x + 8$ (d) $x^2 - 5x - 14 = 0$

(e) $x^2 + 3x - 4 = 0$ (f) $2x^2 - 5x - 3 = 0$

(g) $0 = 3x^2 - x - 4$ (h) $3x^2 - 4x - 4 = 0$

(i) $5x^2 + 13x - 6 = 0$ (j) $0 = 4x^2 - 8x - 5$

(k) $6 + 7x - 3x^2 = 0$ (l) $21 - 11x - 2x^2 = 0$

Here is a page from a pupil's jotter.

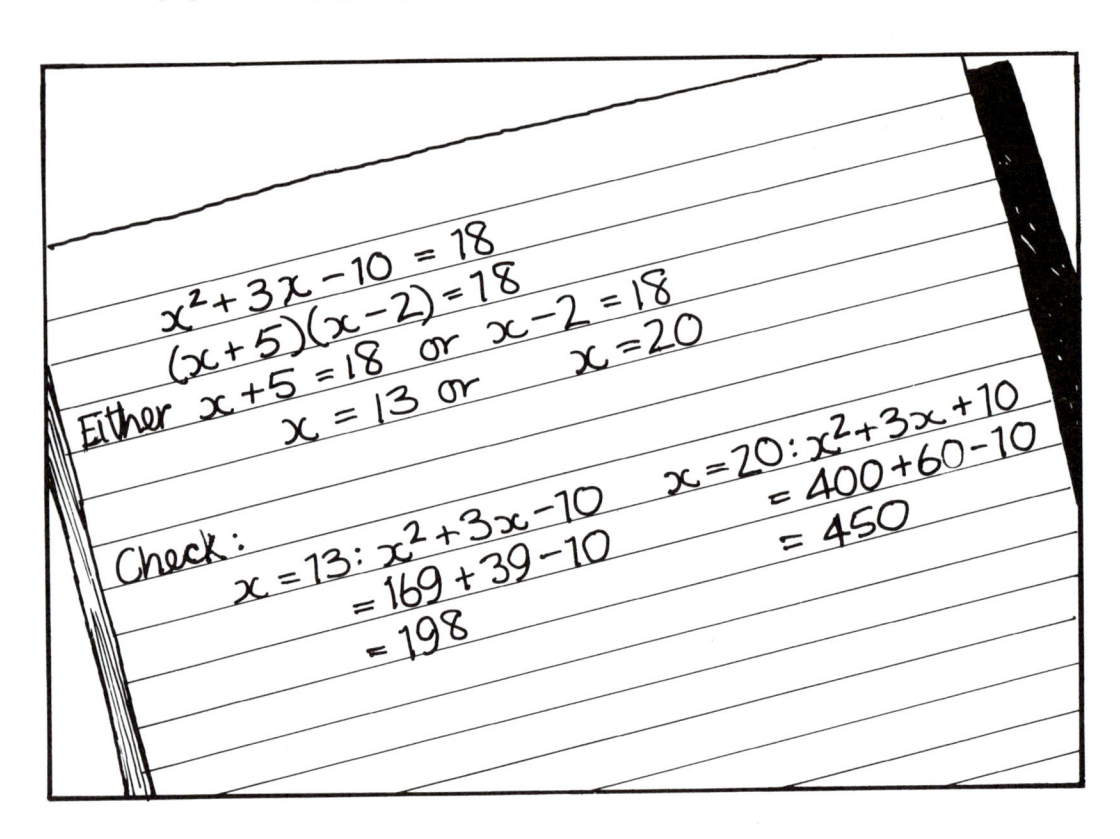

$x^2 + 3x - 10 = 18$
$(x+5)(x-2) = 18$
Either $x + 5 = 18$ or $x - 2 = 18$
$x = 13$ or $x = 20$

Check:
$x = 13: x^2 + 3x - 10$
$= 169 + 39 - 10$
$= 198$

$x = 20: x^2 + 3x + 10$
$= 400 + 60 - 10$
$= 450$

D9 What is the mistake that has been made?

When solving quadratic equations by this method, the first step is to make one side 0. This is usually done by taking all the terms to the other side.

Step 1 Take everything to the left side

$x^2 + 3x - 10 = 18$
$x^2 + 3x - 10 - 18 = 0$
$x^2 + 3x - 28 = 0$ — Collecting like terms

Step 2 Factorise the left side

	−28	28	−14	14	−7	7
	1	−1	2	−2	4	−4
Adding	−27	27	−12	12	−3	**3**

So $x^2 + 3x - 28 = (x + 7)(x - 4)$

Step 3 Rewrite the equation using the factorised form and proceed as before

$x^2 + 3x - 28 = 0$
$(x + 7)(x - 4) = 0$
Either $x + 7 = 0$ or $x - 4 = 0$
$x = -7$ or $x = 4$

Check $x = -7$: $x^2 + 3x - 10$
$= (-7)^2 + 3 \times (-7) - 10$
$= 49 - 21 - 10$
$= 18$

$x = 4$: $x^2 + 3x - 10$
$= 4^2 + 3 \times 4 - 10$
$= 16 + 12 - 10$
$= 18$

D10 Solve these quadratic equations by factorisation. Remember to make one side 0.

(a) $x^2 + 7 = 3x + 61$

(b) $1 = 2x^2 + 3x - 19$

(c) $3(2x^2 + 1) = 5 - x$

(d) $x(x + 3) = 3x^2 + 2(1 + 4x)$

(e) $4x(x - 1) = x - (2 - x)^2$

(f) $(4x - 3)(8x + 5) = 13$

(g) $(x + 3)^2 = 2(x + 6) + 9$

(h) $(x - 3)^2 + (x + 1)^2 = 26$

(i) $(2x + 3)^2 - (x - 1)^2 = 0$

(j) $(4x - 1)(3x + 2) = (x - 5)^2 - 13$

I can't decide which of the three ways to use to solve these equations

Which way is easiest for you?

The factorisation way — and its shorter

So try that way first.

D11 When could a quadratic equation not be solved by the factorisation method?

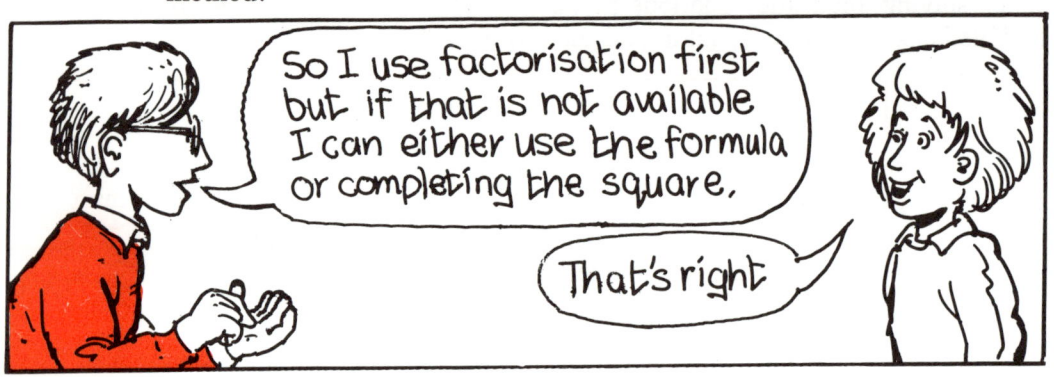

So I use factorisation first but if that is not available I can either use the formula or completing the square.

That's right

D12 Solve these where possible. Where the answers are not exact, give the solutions correct to 3 s.f.

(a) $x^2 - 25 = 0$ — Use the difference of two squares.

(b) $x^2 + 2x - 80 = 0$
(c) $3x^2 + 2x - 2 = 0$
(d) $3(x + 2)^2 = x + 4$
(e) $5x^2 = 3x + 2$
(f) $x^2 = 5(x + 2)$
(g) $4x^2 + 2x = 0$
(h) $5(x + 3) = x^2 - 4$
(i) $2x^2 + x + 1 = 0$
(j) $7 = 2(x + 1)(x + 7) - 100$
(k) $2x^2 + 3x - 5 = 0$
(l) $3x = 4(x^2 - 2)$
(m) $(3 - 2x)(x + 1) = x - 5$
(n) $(x + 3)^2 = (x - 1)(2x + 3)$
(o) $(2x + 1)^2 - (x - 2)(x + 1) = 7$
(p) $(2x + 5)(3x - 1) = (3x - 1)(x + 5) + 11x$

E Using quadratic equations to solve problems

Andrew wants to make a path round two sides of his lawn and has enough materials to cover 62 square metres. His lawn is 16 m long and 13 m wide.

How wide can Andrew make his path so that it is uniform width?

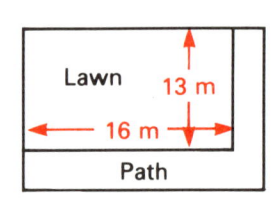

Let x m represent the width of the path.

The total length of the lawn and path is $(16 + x)$ m and the total breadth of the lawn and path is $(13 + x)$ m

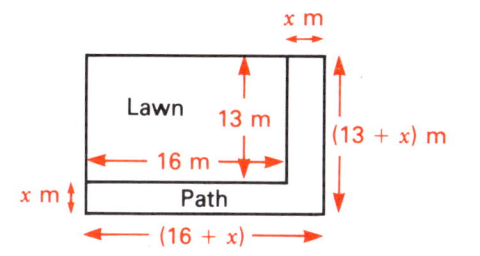

Area of the path = Area of (lawn & path) − Area of lawn
$$62 = (x + 16)(x + 13) - 16 \times 13$$

We can ignore the units as they are all the same.

$$62 = x^2 + 29x + 208 - 208$$
$$0 = x^2 + 29x - 62$$
$$0 = (x + 31)(x - 2)$$
Either $x + 31 = 0$ or $x - 2 = 0$
$x = -31$ or $x = 2$

$x = -31$ is not valid since x is a length and therefore could not be negative.
The path should be 2 m wide.

E1 The length of a rectangular floor is 3 m longer than its breadth. The area of the floor is 270 m².

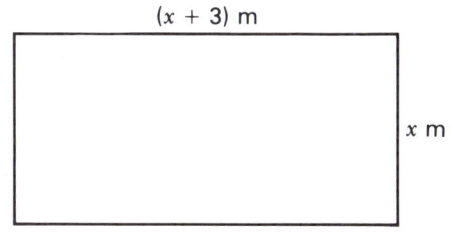

Let x m be the breadth. Form a quadratic equation and hence find the length and breadth of the floor.

E2 If the two rectangles below have the same area, form an equation and solve it.

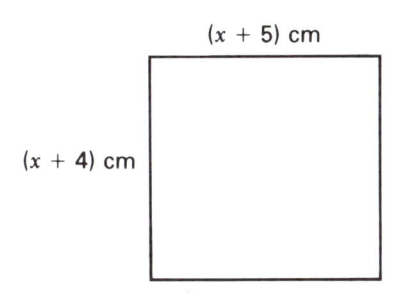

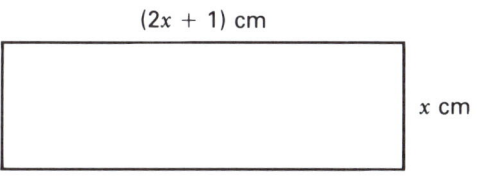

What are the dimensions of each rectangle?

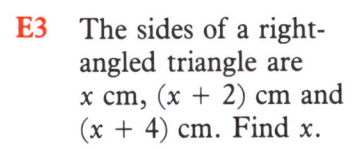

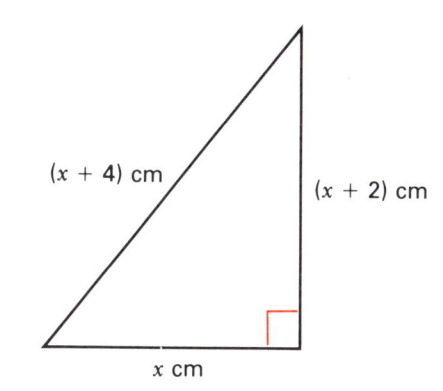

E3 The sides of a right-angled triangle are x cm, $(x + 2)$ cm and $(x + 4)$ cm. Find x.

E4 The sum of the first n numbers of the sequence 4, 10, 16, 22, . . . is $n(3n + 1)$. How many terms add up to 1344?

E5 The squares of two consecutive integers add up to 365. By taking the first integer as n, form a quadratic equation in n and hence find the integers.

E6 Two whole numbers differ by 3. Their product is 154. By taking the first number as n, form a quadratic equation in n and hence find the numbers.

E7 A rope 26 m long is pegged out to form a rectangle whose area is 36 m². By taking one of the sides of the rectangle to be x m, form a quadratic equation in x and hence find the length of the sides of the rectangle.

E8 A sheep pen is made out of 36 m of fencing against a wall. If the area of the pen is 160 m², find the length and breadth of the pen.

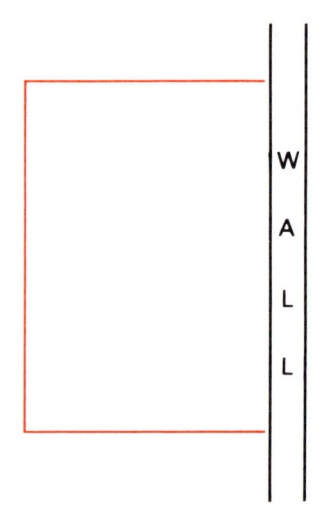

E9 Two consecutive odd numbers are squared and added. The result is 514. What are the numbers?

F Solving equations with fractions

Consider the simple equation

$$\tfrac{3}{4}x + \tfrac{2}{3} = 3\tfrac{1}{3}$$

The easiest way of dealing with equations like this is to clear the fractions. We do this by multiplying the equation by the LCM of the denominators 3 and 4. This number is called the **multiplying factor**.

So

$12 \times \tfrac{3}{4}x + 12 \times \tfrac{2}{3} = 12 \times 3\tfrac{1}{3}$ (Multiplying the equation by 12.)

$9x + 8 = 40$ (The fractions are now cleared and we can proceed as usual.)

$9x = 32$

$x = \tfrac{32}{9}$

F1 Write down the multiplying factor needed for these equations.

(a) $\frac{1}{2} + \frac{3}{4}x = \frac{1}{3}$
(b) $\frac{1}{2}(x + 7) = 4 - \frac{1}{6}x$
(c) $\frac{2}{3} - \frac{1}{2}x = \frac{7}{4}(x - 5)$
(d) $\frac{1}{6}(x + 2) - \frac{1}{2} = 4$
(e) $\frac{5}{3}(x - 2) + \frac{x}{5} = \frac{1}{4}$

F2 Solve these equations by this method.

(a) $\frac{1}{2}x + \frac{3}{4} = 3$
(b) $\frac{2}{5} - \frac{4}{3}y = \frac{9}{10}$
(c) $\frac{2}{3}z - \frac{5}{6} = 5$
(d) $\frac{5}{6}a + 3 = \frac{4}{9}a - 4$
(e) $\frac{2x + 4}{3} = \frac{1}{4}x - 2$
(f) $\frac{y}{3} + \frac{y - 2}{4} = \frac{5}{4}$
(g) $\frac{3z + 2}{6} - \frac{1}{3} = \frac{z + 1}{4}$
(h) $\frac{a + 1}{2} - \frac{a - 7}{5} = \frac{a + 4}{3}$
(i) $\frac{b - 2}{4} - \frac{b + 5}{8} = 2$
(j) $\frac{1}{6}(c + 2) - \frac{1}{8}(c + 25) - \frac{1}{4}(c - 7) = 0$
(k) $\frac{7(x - 3)}{4} - \frac{3(x - 2)}{2} = 8$
(l) $\frac{5(2a - 1)}{6} + \frac{3(a + 2)}{4} = \frac{7(a + 7)}{8} - \frac{5}{6}$

This technique can also be used with inequations. We already know that we can multiply an inequation by a positive number.

F3 Solve these inequations by first clearing the fractions.

(a) $\frac{5}{2}x - 2 > -12$
(b) $\frac{10}{3}a - \frac{2}{3} \leq a - \frac{16}{3}$
(c) $2 + \frac{5}{4}b < 8 + \frac{11}{4}b$
(d) $\frac{3}{5}c - \frac{1}{2}c \geq 1$
(e) $\frac{x - 2}{2} > 2 + \frac{x - 3}{3}$
(f) $\frac{x - 3}{2} < x - 5\frac{1}{2}$
(g) $\frac{9(1 - y)}{4} \geq \frac{6(1 - 2y)}{5}$
(h) $\frac{2}{3}(a + 7) + 12 < 4a$

Remember when multiplying the equation, each denominator should divide (i.e. cancel) with the multiplying factor. If this does not happen, check that you have the correct multiplying factor

Example 1 Solve $\frac{1}{4} = \frac{1}{2} - \frac{1}{x - 2}$ $\frac{1}{4} = \frac{1}{2} - \frac{1}{(x - 2)}$

Put brackets round both terms to keep them together.

123

Multiply the equation by $4(x - 2)$.

$4(x - 2) \times \frac{1}{4} = 4(x - 2) \times \frac{1}{2} - 4(x - 2) \times \frac{1}{(x - 2)}$ *Cancelling*

$x - 2 = 2(x - 2) - 4$
$x - 2 = 2x - 4 - 4$
$x - 2x = -8 + 2$
$-x = -6$
$x = 6$

Example 2   Solve $\dfrac{1}{x - 2} - \dfrac{1}{6} = \dfrac{1}{x + 1}$

$\dfrac{1}{(x - 2)} - \dfrac{1}{6} = \dfrac{1}{(x + 1)}$ *Putting in brackets*

Multiply the equation by $6(x + 1)(x - 2)$

$6(x + 1)(x - 2) \times \dfrac{1}{(x - 2)} - 6(x + 1)(x - 2) \times \dfrac{1}{6} = 6(x + 1)(x - 2) \times \dfrac{1}{(x + 1)}$

$6(x + 1) - (x + 1)(x - 2) = 6(x - 2)$

$6x + 6 - (x^2 - x - 2) = 6x - 12$

$6x + 6 - x^2 + x + 2 = 6x - 12$

$0 = 6x - 12 - 6x - 6 + x^2 - x - 2$

$0 = x^2 - x - 20$ *This is a quadratic equation, so try to factorise first.*

$0 = (x - 5)(x + 4)$

Either $x - 5 = 0$ or $x + 4 = 0$
           $x = 5$ or $x = -4$

The two solutions are -4 and 5

Notice that we can always clear fractions if the multiplying factor is the product of the denominators as in the last examples.

Sometimes this is not the best way to spot the multiplying factor. The best way is to factorise any denominators first so that common factors can be seen more easily. This helps to find the simplest multiplying factor.

F4 Copy and complete this table.

Equation	Equation in factorised form	Multiplying factor
$\dfrac{1}{x-3} - 2 = \dfrac{5x+2}{x^2-2x-3}$	$\dfrac{1}{(x-3)} - 2 = \dfrac{(5x+2)}{(x-3)(x+1)}$	
$\dfrac{3x^2}{x^2-1} = \dfrac{4}{x-1}$		
$\dfrac{5}{x-2} = \dfrac{3}{x} + \dfrac{12}{x^2-2x}$		
$\dfrac{x}{3x-6} + \dfrac{x}{2x-4} = 7$		
$\dfrac{4}{2y-3} = \dfrac{5}{y} + \dfrac{3}{2y^2-3y}$		
$\dfrac{9}{(3x+2)^2} = \dfrac{4}{3x+2} - \dfrac{1}{12x+8}$		

F5 Solve these equations

(a) $\dfrac{4}{x} + 3 = \dfrac{5}{x-1}$

(b) $6 - \dfrac{2}{1-a} = \dfrac{1}{a}$

(c) $\dfrac{3a+8}{a+2} - \dfrac{2a+7}{a+4} = 1$

(d) $\dfrac{1}{a-2} - \dfrac{3}{a^2+a-6} = \dfrac{1}{2}$

(e) $\dfrac{1}{a^2-1} - \dfrac{1}{a-1} = \dfrac{6}{5}$

(f) $\dfrac{3}{4} - \dfrac{1}{a+1} = \dfrac{1}{a^2+2a+1}$

(g) $1 + \dfrac{1}{x-3} = \dfrac{x+10}{x^2-9}$

(h) $\dfrac{3}{x^2+4x+3} = \dfrac{2}{5} - \dfrac{1}{x+3}$

G More problems involving quadratic equations

G1 The number of diagonals, d, of a polygon is given by the formula

$$d = \tfrac{1}{2}n(n-3)$$

where n is the number of sides of the polygon.

(a) How many sides has the polygon with 90 diagonals?

(b) Is there a polygon with 50 diagonals? Explain your answer.

G2 (a) What is the reciprocal of $\tfrac{2}{3}$?

(b) The difference between a real number and its reciprocal is $\tfrac{21}{10}$. By taking x to be the number, form an equation and solve it.

(c) Explain the solution to the equation.

G3 Mrs T. Esbee is going to invest £6000 in a special two-year bank account. She needs her £6000 to grow to £6615 by the end of the two years and tries to work out what interest rate she must find to achieve this.

The formula

$$A = P(1 + \tfrac{r}{200})^2$$

where A stands for the amount of money in the bank at the end of the two years, P stands for the original principal placed in the account and r stands for the interest rate in % can be used in cases like this.

By forming an equation, find what interest rate Mrs Esbee must find to achieve the required return on her money.

G4 The point $(a, 8\tfrac{1}{2})$ lies on the curve $y = x + \tfrac{4}{x}$. Find the possible values of a.

G5 A circle has equation $x^2 + y^2 - 10x - 6y + 9 = 0$.
Calculate the coordinates of the points where it crosses the x- and y-axes.

Money management 8: Time for a holiday!

A Foreign exchange

Tables like this tell you the amount of foreign currency you get for every £1. This is called an **exchange rate**.

Exchange rates vary daily. Look for a recent table in a daily newspaper and see how it compares with the one above.

```
┌─────── £ TODAY ───────┐
      Your Holiday £
  AUSTRIA.......21.01 schillings
  FRANCE..........10.09 francs
  GERMANY.........2.98 marks
  GREECE.........262 drachmas
  IRELAND..........1.125 punts
  ITALY................2160 lire
  PORTUGAL.......252 escudos
  SPAIN..........187.00 pesetas
  SWISS.............2.59 francs
  U.S...............1.607 dollars
```

A1 Use the newspaper to copy and complete this table:

Currency	Exchange rate
French francs	
US dollars	
Pesetas	
Swiss francs	
Lire	
Drachmas	

A2 Change the prices below into the currencies shown

  (i) using the table printed above

  (ii) using your own table (question A1)

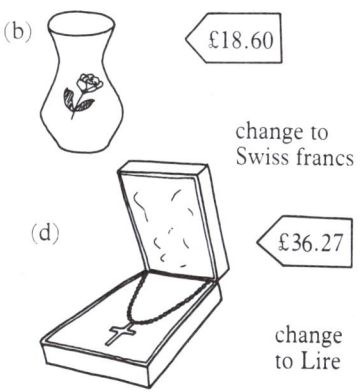

(a) £7.75 — change to US dollars
£1 = $1.607
£7.75 = 7.75 × $1.607
     = $ . . .

(b) £18.60 — change to Swiss francs

(c) £6.25 — change to French francs

(d) £36.27 — change to Lire

A3 The following objects were bought abroad. Work out how much they would cost in £, again using both exchange rates.

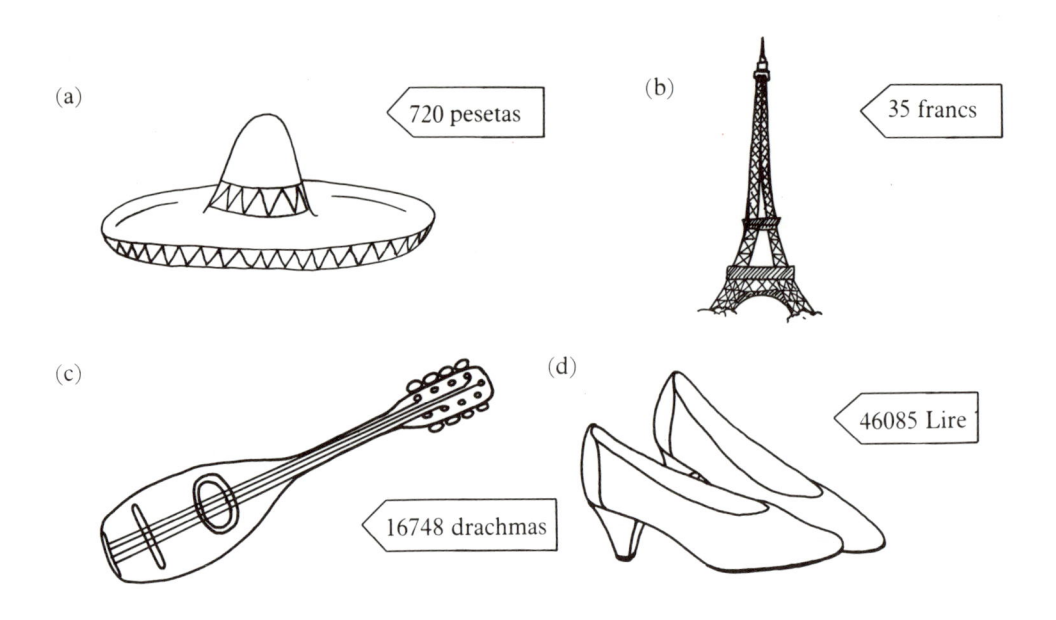

(a) 720 pesetas

(b) 35 francs

(c) 16748 drachmas

(d) 46085 Lire

A4 These items were bought abroad by foreign visitors. Change each of the prices to the currency shown, again using both tables.

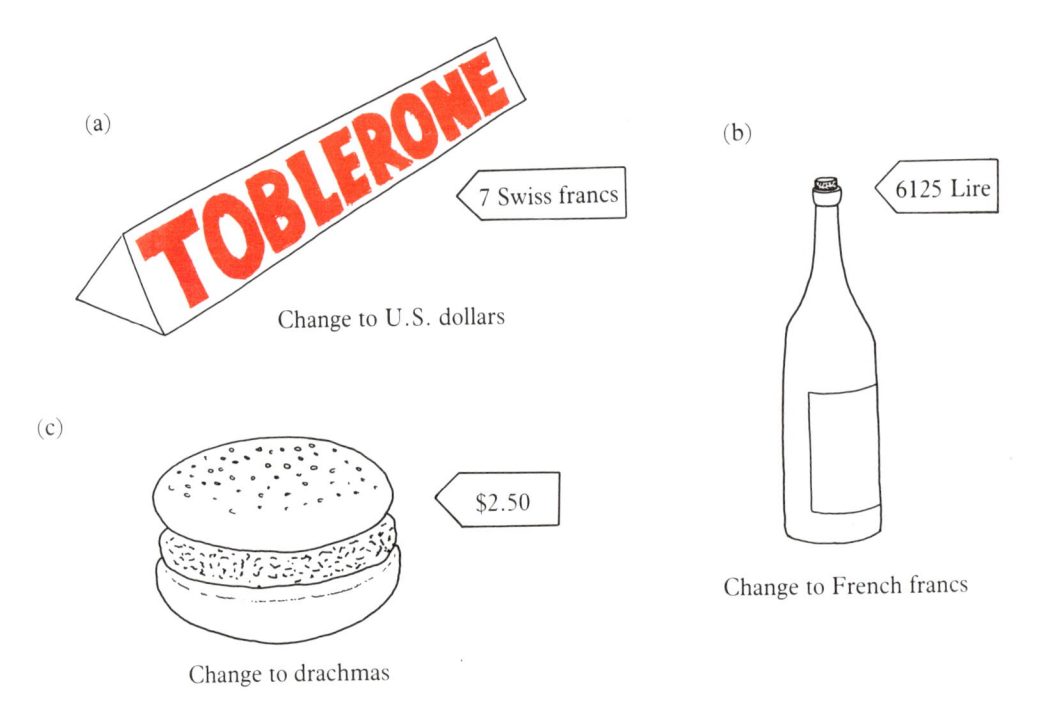

(a) 7 Swiss francs

Change to U.S. dollars

(b) 6125 Lire

Change to French francs

(c) $2.50

Change to drachmas

B The plane time

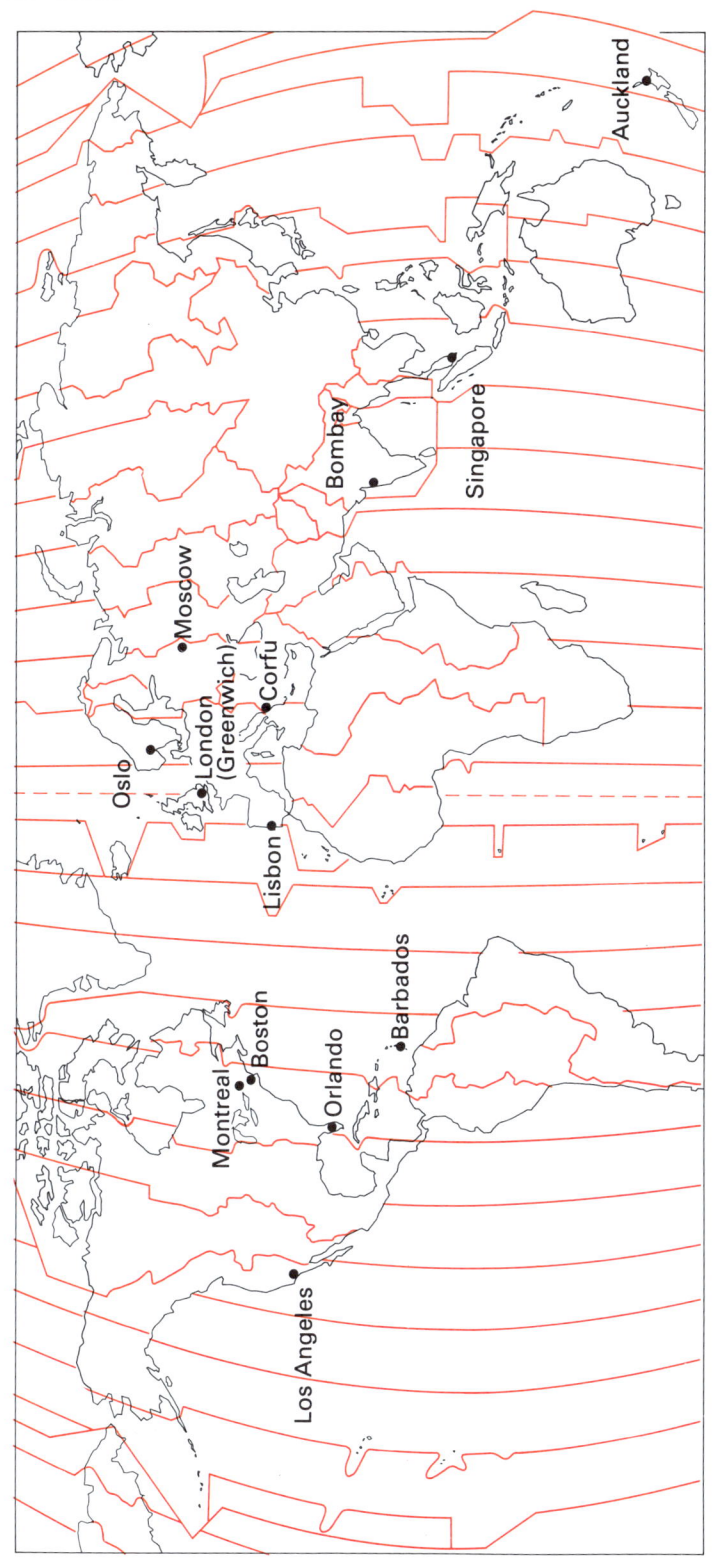

Standard time is different in different parts of the world. For example, when it is daytime in Britain it is night time in Australia.

The number of hours by which the time in another country varies from that in Britain is called the **time difference**. It is measured from the Royal Observatory at Greenwich (Greenwich Mean Time).

Places **east** of Greenwich are **ahead** in time.
Places **west** of Greenwich are **behind** in time.

In spring we put our clocks forward an hour for British Summer Time. We put them back an hour in autumn. In this Section we shall ignore British Summer Time. Our calculations will be valid for winter only.

For example, Oslo is 1 hour ahead.

Time at Greenwich
Midday

Time in Oslo
1300

Montreal is 5 hours behind.

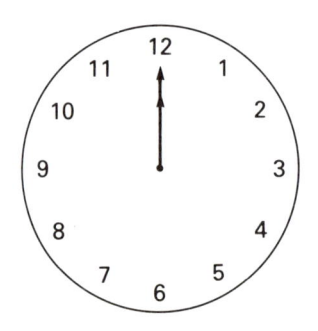

Time at Greenwich
Midday

Time in Montreal
0700

B1 If it is midday in Greenwich, what time is it in

(a) Los Angeles (b) Moscow (c) Auckland (d) Lisbon?

B2 Linsey is flying from London to Boston. The flight takes 7 hours.

(a) If she took off from London at 3 p.m. and did not change her watch during the flight, what time would her watch show when she arrived in Boston?

(b) What is the time difference between London and Boston?

(c) What is the time in Boston when Linsey arrived?

B3 Here is a table containing the departure times (from Gatwick) and local arrival times of a number of flights.

Work out for each flight, the actual flying time. The first one has been done for you.

Destination	Departure	Arrival	Flight time
Los Angeles	10 55	15 40	12 h 45 min
Orlando	11 30	15 30	
Barbados	12 10	18 50	
Bombay	18 00	08 10	
Singapore	14 30	13 30	

C On holiday

Cathy and Colin are going on holiday to the Greek island of Corfu. They will be staying in the Royal Hotel in Kanoni.

Here is some information about the resort and the hotel.

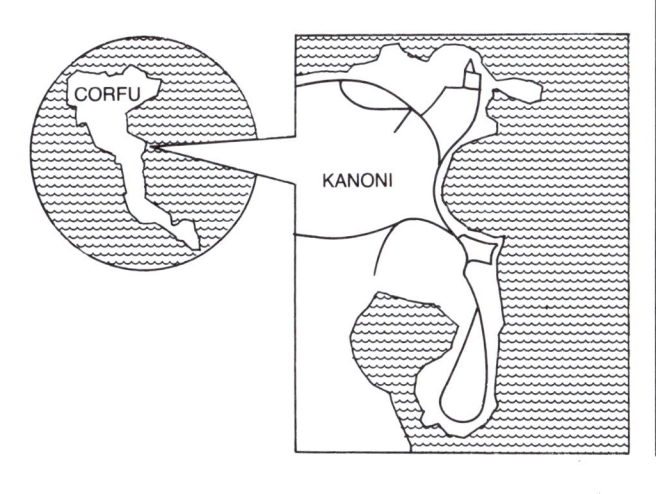

Accommodation	Royal		
Board	Bed & Breakfast		
Accommodation code	SKR		
Brochure page	250		
Flights available	All flights		As Child Room
Apartment type	–		
Number sharing	–		
Number of nights	7	14	
Departures on or between 1 May–12 May	165	205	50
13 May–23 May	175	224	50
24 May–30 May	189	251	40
31 May–16 June	185	241	40
17 June–23 June	191	236	30
24 June–3 July	185	235	30
4 July–14 July	207	262	30
15 July–21 July	194	251	30
22 July–4 Aug	230	307	25
5 Aug–21 Aug	218	307	25
22 Aug–28 Aug	226	269	25
29 Aug–5 Sep	235	281	30
6 Sep–11 Sep	225	275	30
12 Sep–18 Sep	223	271	30
19 Sep–25 Sep	195	235	30
26 Sep–17 Oct	178	225	50
18 Oct–31 Oct	191	–	50
Supplements per person per night	Balcony 70p Single Room £3.40		

FLIGHT INFORMATION
You can fly to CORFU from:

GATWICK	3¼HRS	BIRMINGHAM	3½HRS
LUTON	3¼HRS	MANCHESTER	3¾HRS
BRISTOL	3½HRS	NEWCASTLE	3¾HRS
E.MIDLANDS	3½HRS	GLASGOW	4HRS

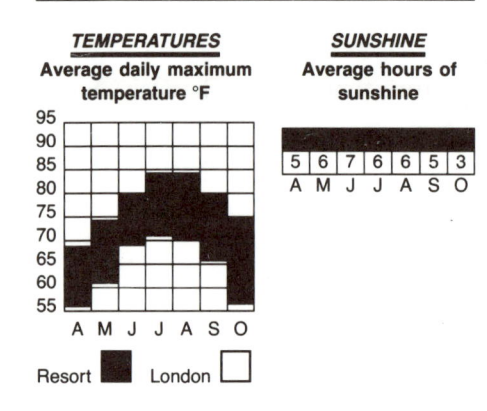

TEMPERATURES — Average daily maximum temperature °F

SUNSHINE — Average hours of sunshine

5	6	7	6	6	5	3
A	M	J	J	A	S	O

Resort ■   London □

FLIGHTS
FLIGHTS TO CORFU

Flight Code	Depart Day	Time	Return Day	Time	Flight Suppl.
From Gatwick					
9487	Wed	10.15	Wed	17.45	£0*
9489	Sat	07.15	Sat	14.30	£15*
9486	Sat	20.30	Sun	03.45	£9*
From Stansted					
9488	Sat	08.15	Sat	23.45	£15
9473	Sat	22.15	Sun	05.30	£0
From Luton					
9490	Sat	20.45	Sun	04.00	£12*
From Bristol					
9491	Sat	21.00	Sun	04.45	£19
From Birmingham					
9492	Wed	14.30	Wed	22.15	£14
From East Midlands					
9493	Wed	15.15	Wed	22.45	£16
From Manchester					
9495	Wed	23.30	Thu	07.15	£9*
9494	Wed	08.00	Wed	15.45	£20*
9496	Sat	07.45	Sat	15.30	£34*
From Newcastle					
9498	Wed	21.15	Thu	05.30	£23
†From Glasgow					
9497	Sat	07.00	Sat	15.45	£49

†Glasgow departures: Add £15 on all holiday lengths 16th, 23rd July.
†Glasgow departures: Reduction of £5 on all holiday lengths 28th May.
*Boeing 767 flights planned.
All Corfu prices are for holidays on the flight with no supplement. Not all hotels and apartments are available on this flight.

C1 If Cathy and Colin wish to start their fortnight's holiday on July 16th, how much will it cost each of them?

C2 If they want a room with a balcony, how much extra will each of them have to pay?

C3 How much extra will each of them have to pay to fly from Glasgow?

C4 Calculate the total cost of their fortnight's holiday at the Royal Hotel in a room with a balcony, flying from Glasgow?

C5 If their plane leaves Glasgow at 08 00, work out their local landing time on Corfu.

C6 Which month of the year has the highest average temperature and average hours of sunlight?

C7 Can you suggest something that may put Cathy and Colin off their hotel?

C8 Cathy is taking £225 spending money, while Colin is taking £275.

(a) How much, in local currency, will each of them receive? (Use the table on page 127.)

(b) Use your table (question A1) to work out their spending money.

(c) Work out the differences between (a) and (b) as percentages.

11 Linear programming

A Splitting the x, y plane

A1 The diagram shows the line $x = 4$.
It consists of all points whose x-coordinate is 4.

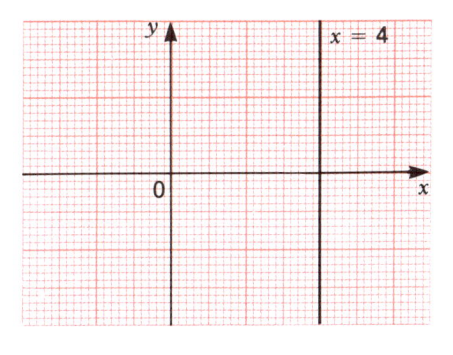

(a) Copy the diagram onto squared paper and plot with a dot the points:

$(5, 7), (7, -2), (4\frac{1}{2}, 1), (5, -4), (6, 3), (8, 0)$

(b) Where do all these points lie?
What do you notice about their coordinates (both of them)?

(c) On the same diagram plot with a cross the points:

$(3, 1), (0, -1), (2\frac{1}{4}, -5), (-3, 2), (0, 4), (-2, -5)$

(d) What do you notice about the coordinates of the points marked with a cross?

A2 (a) Draw axes numbered from -6 to 6 on squared paper. Draw the line with equation $y = -2$.

(b) What do you know about the coordinates of points on this line?

(c) Plot the points: $(5, -1), (-2, 0), (-4, 5), (3, 4), (-3, -1\frac{1}{2})$, $(0, -1)$.

Write down an inequation which is true for all these points.

(d) Write down an inequation which is true for all points below the line. Choose four points below the line and check that they satisfy the inequation.

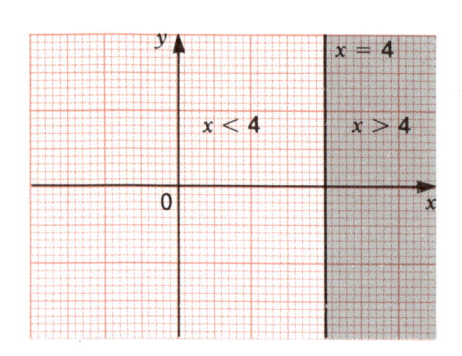

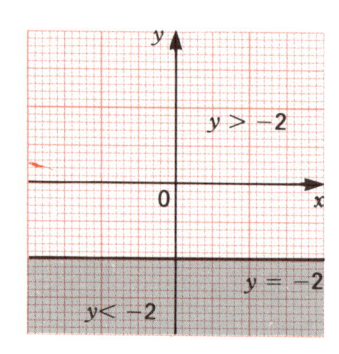

These lines split the x, y plane into two regions.
Each region is described by an inequation which is closely linked to the equation of the line.

A3 What lines would be drawn to show these regions?

(a) $x > 5$ (b) $y > -1$ (c) $x < -2$ (d) $y < 2$

(e) $x > 3\frac{1}{2}$ (f) $y < -\frac{1}{2}$ (g) $x > 0$ (h) $y < -5$

A4 What inequations would give these regions? (The regions are the **unshaded** parts.)

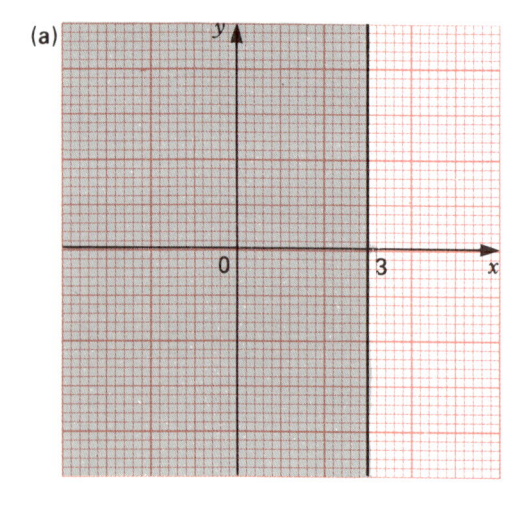

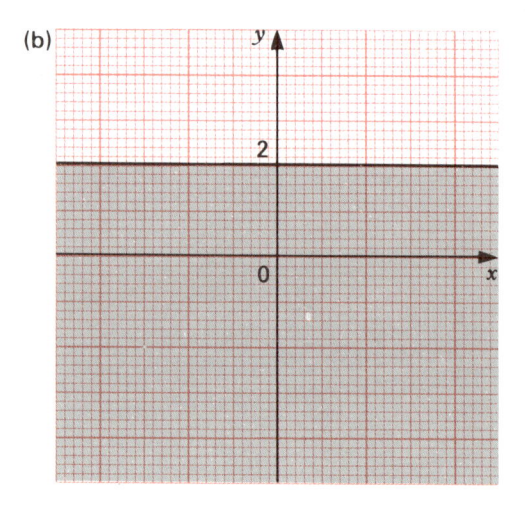

(c)

(d)

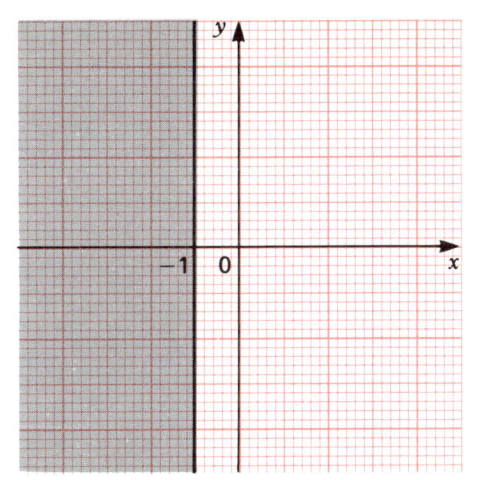

(e)

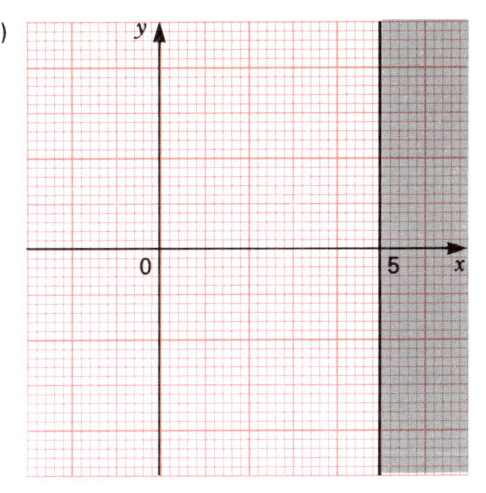

(f)

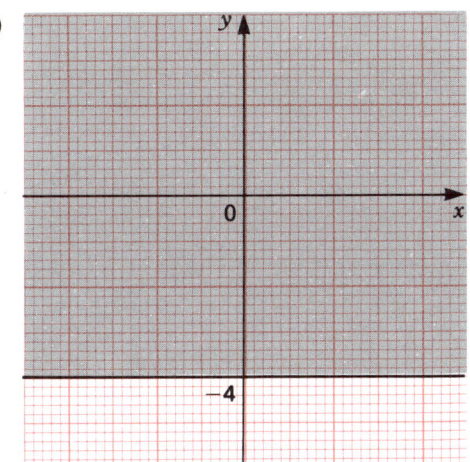

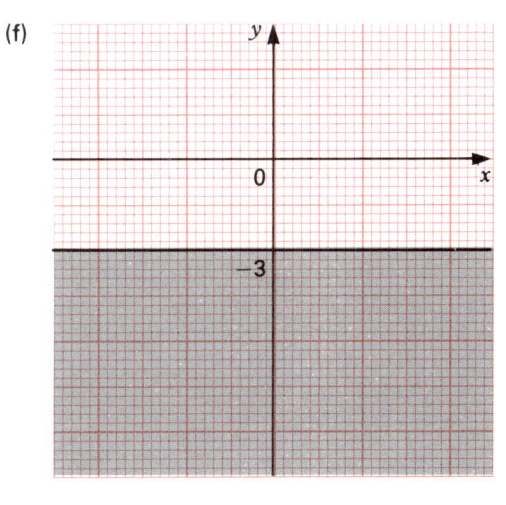

A5 Draw rough sketches to show where these regions lie. (Remember to leave the region unshaded.)

(a) $x < 1$ (b) $x > 7$ (c) $y > 2$ (d) $y > -3$
(e) $x > -3$ (f) $y < -2$

A6 Describe in words how to find the regions given by

(a) $x < a$, if a is negative (b) $y > a$, if a is positive
(c) $x > a$, if a is positive

All of the lines involved so far have been of the form $x = a$ or $y = a$ where a is a number.

We have met lines of the form $y = ax$ (like $y = 2x$, $y = -5x$ etc.) and $y = ax + b$ (like $y = 3x - 1$, $y = -x + 7$ . . .) and they too divide the x, y plane into two regions.

Consider the line $2x - y = 8$

We can draw its graph by finding the points where it cuts the axes.

Let $x = 0$: $0 - y = 8$
$\ -y = 8$
$y = -8$
So the line crosses the y-axis at $(0, -8)$.

Let $y = 0$: $2x - 0 = 8$
$2x = 8$
$x = 4$
So the line crosses the x-axis at $(4, 0)$.

The red numbers below show the value of $2x - y$ at different points

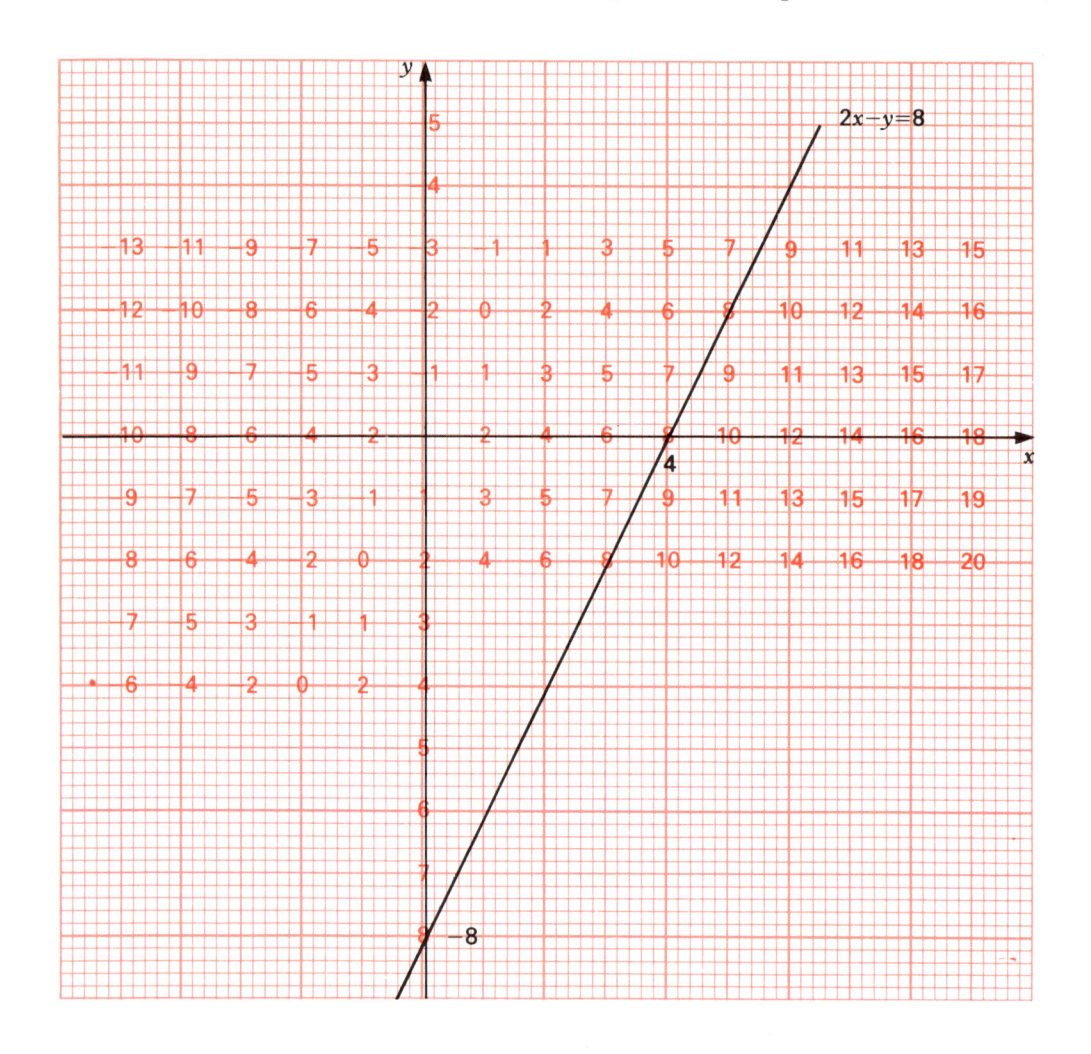

A7 (a) What do you notice about the value of $2x - y$ for points **above** the line?

(b) Give the inequation that describes the region above the line.

A8 (a) Draw the line $3x + 2y = 12$ on squared paper.

136

(b) Work out the value of $3x + 2y$ at the points where two grid lines cross.

(c) What do you notice about the value of $3x + 2y$ for points **above** the line?

(d) Give the inequation that describes the region above the line.

These last two examples illustrate that above the line can be an inequation with a $>$ sign or a $<$ sign.

Suppose you are given the line $3x - y = 6$ and want to decide which side of the line is $3x - y > 6$.

Remember that for **all** points on one side of the line, $3x - y > 6$; on the other side, $3x - y < 6$. So take a point on one side of the line (it doesn't matter which side) and work out $3x - y$.

If the answer is more than 6, then that side is
$$3x - y > 6$$

If the answer is less then 6, then that side is
$$3x - y < 6$$

For example, take the point (4, 1). Now for (4, 1), $3x - y = 3 \times 4 - 1 = 11$. Since $11 > 6$, the side below the line is
$$3x - y > 6$$
(and above the line must be $3x - y < 6$)

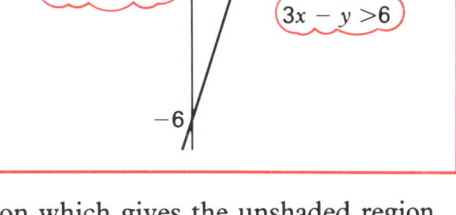

A9 Use this method to find an inequation which gives the unshaded region in each of these sketches.

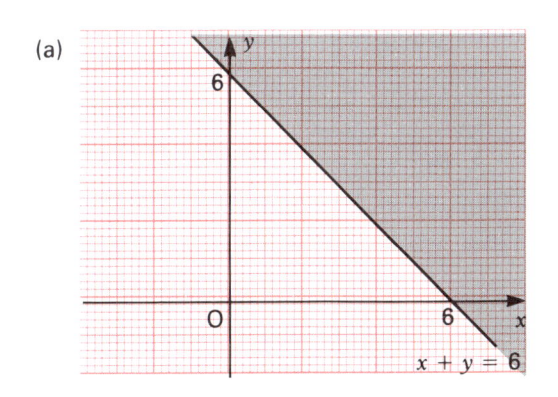

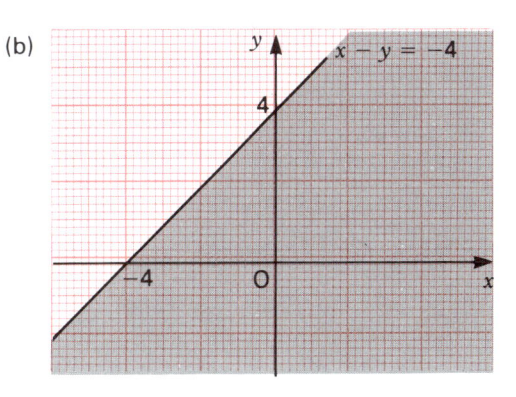

(a) line $x + y = 6$, shaded region above the line

(b) line $x - y = -4$, shaded region below the line

137

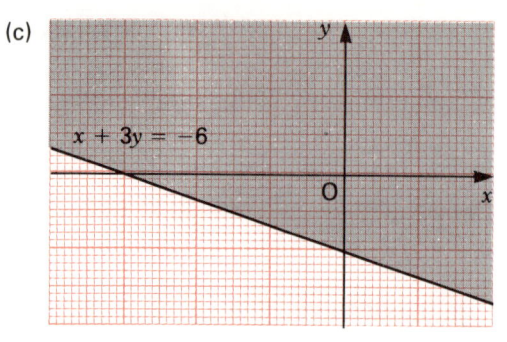

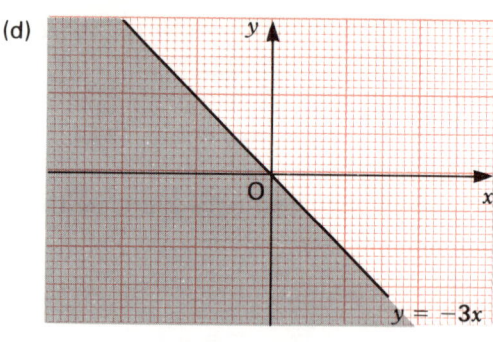

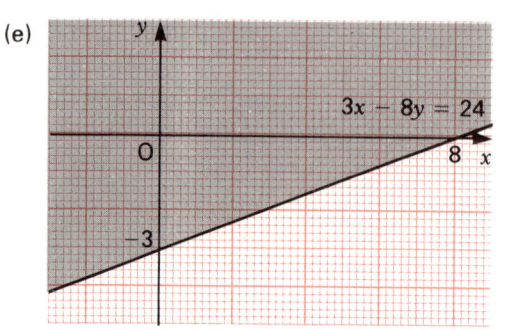

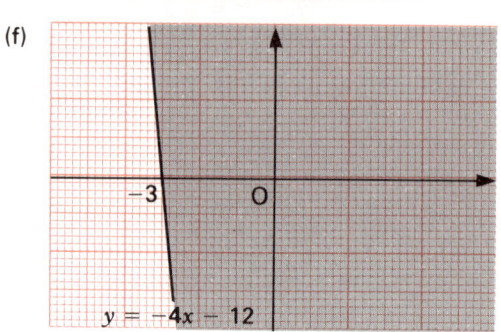

(c) [graph: $x + 3y = -6$]

(d) [graph: $y = -3x$]

(e) [graph: $3x - 8y = 24$]

(f) [graph: $y = -4x - 12$]

B Boundary lines

Consider the graph of $2x + y = 8$ which is sketched below.

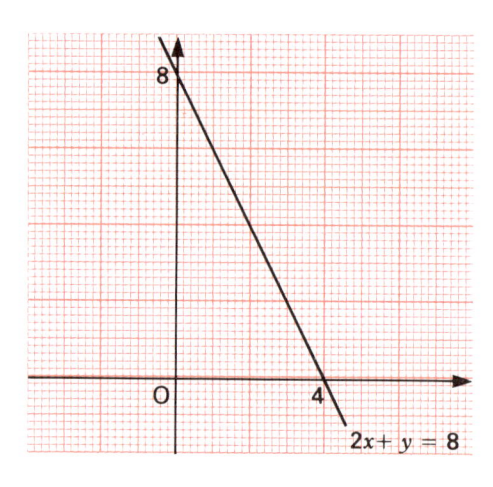

[graph: $2x + y = 8$]

We now have a technique for finding an inequation that describes one side of the line. So we can find $2x + y > 8$ and $2x + y < 8$. But there are two other inequations we can make with the terms of the line.

Now $2x + y \geqslant 8$ means $\qquad$ $2x + y > 8$ $\qquad$ or $\qquad$ $2x + y = 8$

⟶ These are points on one side of the line $2x + y = 8$.

⟶ These are points on the line $2x + y = 8$.

138

So this inequation describes one side of the line and also **includes the line itself**.

To show this on a graph we use this convention:

If the line is **not** included, then draw a **broken** line.
If the line **is** included, then draw a **normal** line.

Of course, we will still leave the region unshaded.

Examples

Show the region described by $x + 4y < 8$.

First draw the line $x + 4y = 8$.

Let $x = 0$: $0 + 4y = 8$
$4y = 8$
$y = 2$

So the line crosses the y-axis at $(0, 2)$.

Let $y = 0$: $x + 0 = 8$
$x = 8$

So the line crosses the x-axis at $(8, 0)$.

Since the line is not included, draw a broken line.

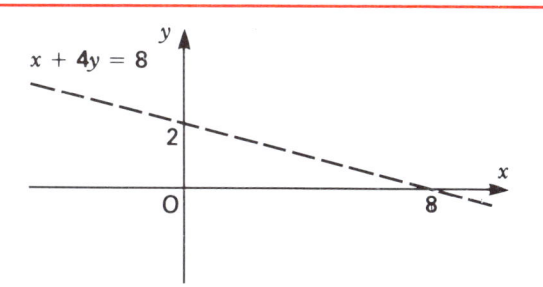

Take a point on one side of the line, say $(0, 0)$.

For $(0, 0)$: $x + 4y = 0 + 0 = 0$

Since $0 < 8$, this means that $x + 4y < 8$ is below the line. Leave the region unshaded.

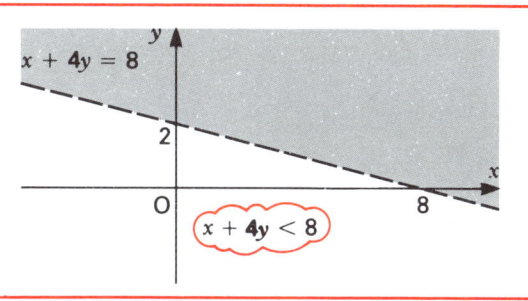

B1 Use the working in the example above to draw the region given by $x + 4y \geqslant 8$.

B2 On separate diagrams, draw the regions described by these inequations.

(a) $x + y > 3$

(b) $2x - y \leqslant 6$

(c) $3x + y < -12$

(d) $4x - 2y \geqslant 20$

(e) $y \geqslant 4x$

(f) $y > 2x - 4$

(g) $2x + y \leqslant -4$

(h) $x < 3y + 12$

(i) $5x + 2y > -10$

(j) $-3x + y \geqslant 9$

C Regions from more than one inequation

The region $x + y > 6$ is above the line a but not on it.

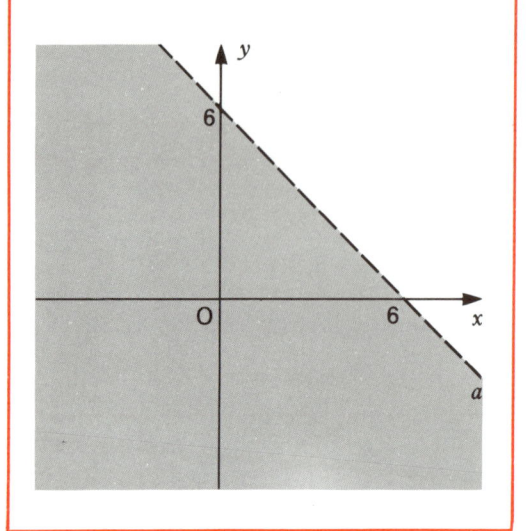

The region $x - 2y < -8$ is above the line b but not on it.

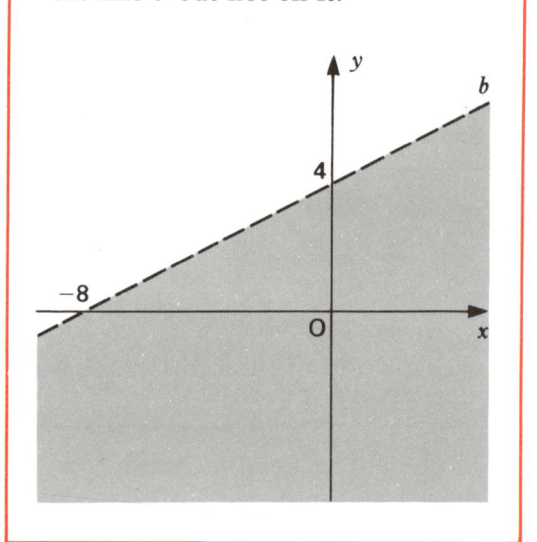

The region which is above line a and above line b can be found by placing one diagram over the other.

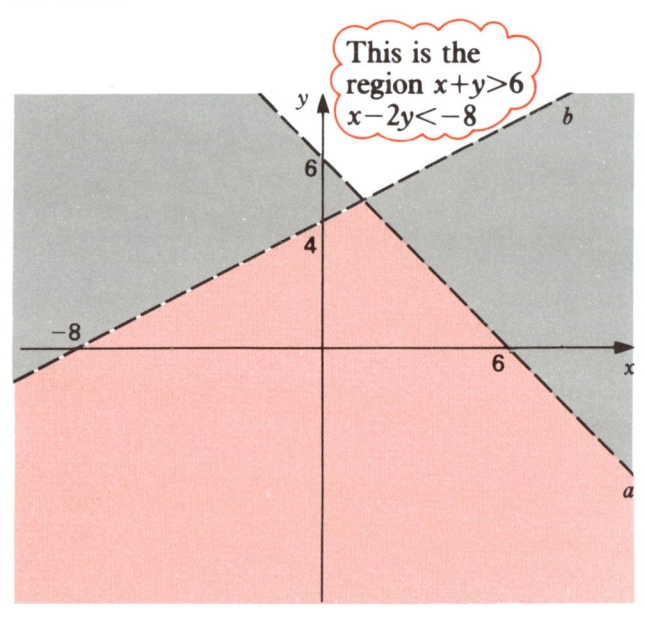

This is the region $x+y>6$
$x-2y<-8$

C1 Take a point in the unshaded region.

(a) Does it satisfy
   $y > 2x$ or
   $y < 2x$?

(b) Does it satisfy
   $x + 2y > 10$ or
   $x + 2y < 10$?

(c) Write down two inequations that describe the unshaded region.

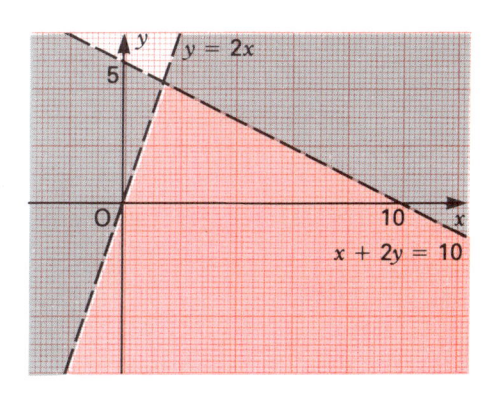

C2 Write down two inequations that describe each of these regions.

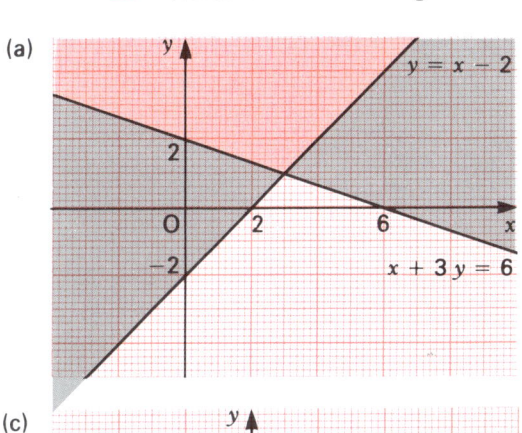

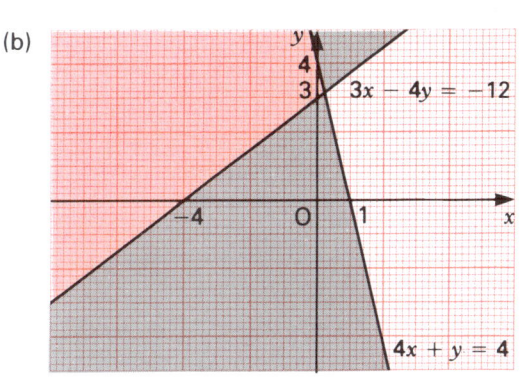

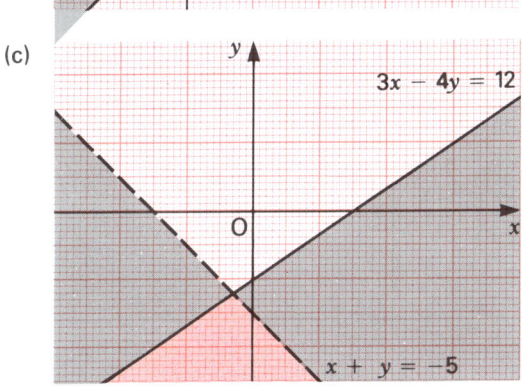

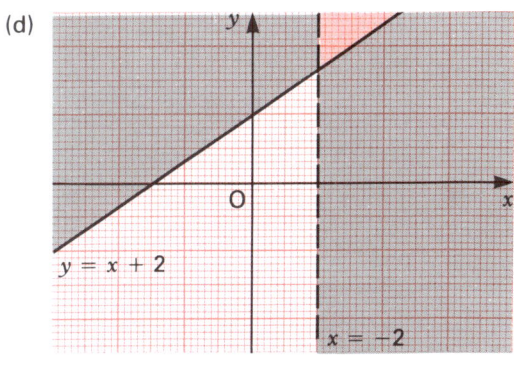

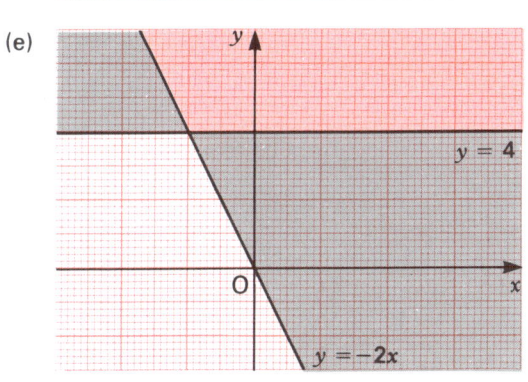

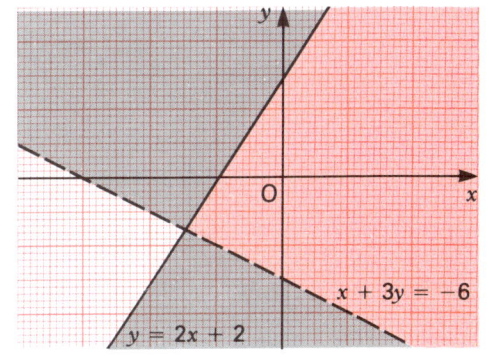

C3 Draw the regions described by the following inequations.

(a) $x + y \geq 6$ and $2x - y \leq 4$

(b) $3x + y < 6$ and $x - y < 2$

(c) $y \geq 3x + 12$ and $x + 2y < 8$

(d) $x > 2y - 10$ and $2x + 3y \leq 12$

(e) $x + 4y \geq 8$ and $3x - 5y \leq 15$

(f) $x \geq 4$ and $x + y \leq 10$

(g) $y \leq 0$ and $3x - y \geq 6$

D Using inequations to solve real problems

Audrey is making two kinds of jumper.

Cardigans need 500 g of wool and take 6 hours to make.
Pullovers need 400 g of wool and take 9 hours to make.

She has 2 kg of wool and 36 hours of time.

If she sells the cardigans for £12 and the pullovers for £10, how many of each should she knit to make as much money as she can?

The two commodities required here are wool and time.
They are both limited.

Audrey can use up to 2 kg (2000 g) of wool and up to 36 hours of time.

Make a table showing all the given information.

	Cardigans	Pullovers	Available
Wool	500 g	400 g	2000 g
Time	6 h	9 h	36 h

Suppose Audrey makes x cardigans and y pullovers.

Then $x \geq 0$ and $y \geq 0$ and both must be whole numbers.

> Audrey does not want to make half a pullover!

Consider one of the commodities.

Wool 1 cardigan needs 500 g.   1 pullover needs 400 g.
        x cardigans need $500x$ g. y pullovers need $400y$ g.

The total amount of wool needed to make x cardigans and y pullovers is

$$(500x + 400y) \text{ g}$$

and this cannot be more than 2000 g.

So $500x + 400y \leq 2000$

> Divide by 100

> The = sign is there since Audrey could use **all** her wool.

So $5x + 4y \leq 20$

Time 1 cardigan needs 6 h.   1 pullover needs 9 h.
        x cardigans need $6x$ h. y pullovers need $9y$.

The total amount of time needed to make x cardigans and y pullovers is

$$(6x + 9y) \text{ hours}$$

and this cannot be more than 36 hours.

So $6x + 9y \leq 36$

> Divide by 3

So $2x + 3y \leq 12$

Draw a diagram showing the region described by $5x + 4y \leq 20$ (the wool limit) and $2x + 3y \leq 12$ (the time limit)

> These points give the number of cardigans (x) and the number of pullovers (y) that are possible

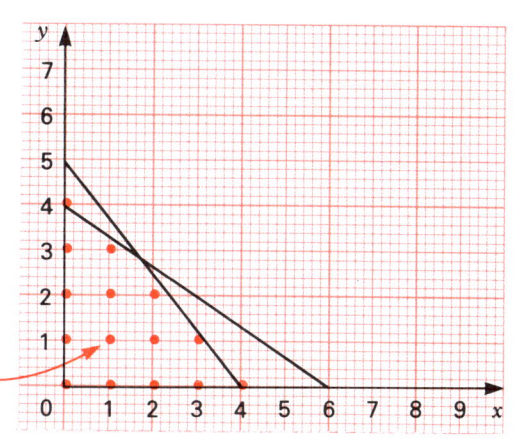

On 1 cardigan Audrey makes £12.
On 1 pullover Audrey makes £10.
On x cardigans Audrey makes £$12x$.   On y pullovers Audrey makes £$10y$.

So Audrey will make £(12x + 10y) altogether.

She wants to find the point which will give her most money.

Here is a diagram showing the value of 12x + 10y at each of the points in the region.

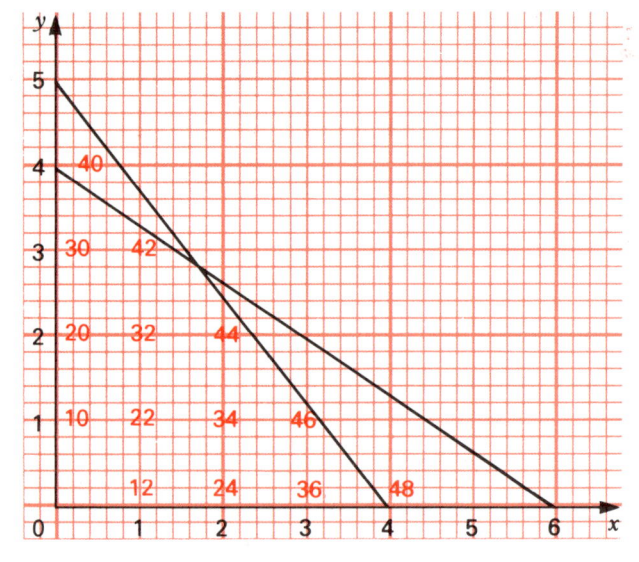

The largest amount of money is £48, at the point (4, 0).
This means that Audrey should make 4 cardigans and no pullovers and she will make £48.

D1 Mrs Vassel owns a sweet shop. She has bought 12 kg of milk chocolates and 18 kg of plain chocolates to make up mixed boxes which she will call 'Half & Half' and 'Dark Secret'.

Each 500 g box of 'Half & Half' contains 250 g of milk and 250 g of plain chocolates, and a 500 g box of 'Dark Secret' contains 100 g of milk and 400 g of plain chocolates.

If the profit on a box of 'Half & Half' is £1·50 and on 'Dark Secret' is £2, find how many boxes of each kind Mrs Vassel should sell to make the most profit.

D2 Andrew is a market gardener and grows carrots and cabbages. He has 24 gardeners who work for him. The cost of planting one hectare of carrots is £5 and of cabbages is £8.

If 6 gardeners are required to plant a hectare of carrots and 4 gardeners required to plant a hectare of cabbages, find the possible different areas Andrew can plant given that he is allowed £40 for planting.

What is the largest area he can plant?

D3 Mrs Growem is another market gardener who also grows carrots and cabbages. She only has 6 hectares of land but does not have a limit on what she is allowed to spend on planting.

Her work force consists of 30 gardeners, and she allows 6 workers for each hectare of carrots and 4 for each hectare of cabbages, the same as Andrew.

If Mrs Growem makes £60 profit per hectare of carrots and £50 profit per hectare of cabbages, find how many hectares of each she should plant to make maximum profit.

D4 Morag does two exercises each day – sit-ups and star jumps. She estimates that she can do 12 sit-ups per minute and 8 star jumps per minute. She must do at least 24 exercises each day and to encourage her to do this she gets 'points'. For each sit-up she gets 10 points and for each star jump she gets 20 points. Given that she must do her exercises in complete minutes, how few sit-ups and star jumps can Morag get away with and still get at least 40 points?

12 Graphs 3

A Rough sketches of quadratic functions

In *Mathematics for Credit 1* we found the roots of a quadratic function from its graph.

We can now **calculate** these roots.
Example
Calculate the roots of $y = x^2 + 4x - 12$.

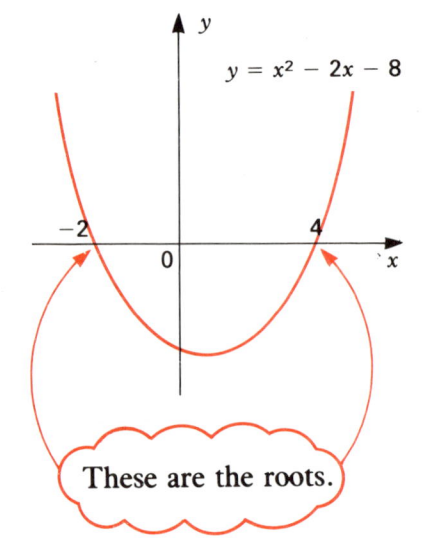

$y = x^2 - 2x - 8$

These are the roots.

The roots are the points where the graph cuts the x-axis and so the y-coordinates of these points are zero. So to find the roots we substitute $y = 0$ into the equation of the curve.

The roots are given by
$x^2 + 4x - 12 = 0$

This is a quadratic equation. We will try to factorise it to solve it.

$(x + 6)(x - 2) = 0$
Either $x + 6 = 0$ or $x - 2 = 0$
$x = -6$ or $x = 2$
The roots are -6 and 2.

This helps us to make a **rough sketch** of the function.
A rough sketch is not as accurate as a graph but it gives an idea of the shape of the graph.

Since the coefficient of the x^2 term is positive, the graph is U-shaped. The rough sketch of the graph is

From the symmetry of the graph, the x-coordinate of the turning point is -2. By substitution, the y-coordinate is given by

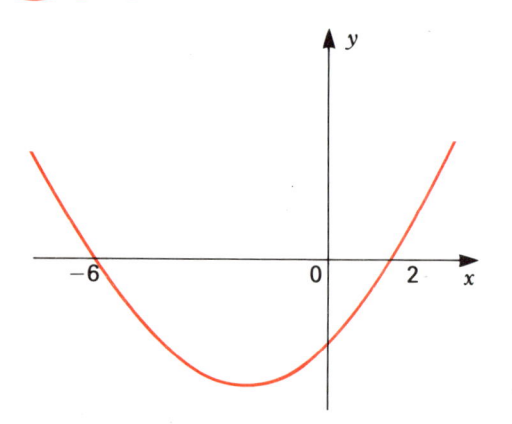

If the coefficient is positive, the graph is ∪-shaped.
If the coefficient is negative, the graph is ∩-shaped.

146

$y = (-2)^2 + 4 \times (-2) - 12 =$
$4 - 8 - 12 = -16$

If you wish to check your rough sketch, find another point on the graph, for example, the point where it cuts the y-axis (when $x = 0$).

A1 Calculate the roots of these quadratic functions. Then make a rough sketch of each one on separate diagrams, marking in the coordinates of the turning point.

(a) $y = x^2 + 5x - 14$
(b) $y = x^2 + 5x + 4$
(c) $y = -x^2 - 3x + 4$
(d) $y = -x^2 + 10x - 16$
(e) $y = 2x^2 - 11x + 5$
(f) $y = -3x^2 + x + 2$
(g) $y = x^2 - 9$
(h) $y = -6x^2 - 37x - 45$

A2 (a) Try this technique with $y = x^2 + 4x + 4$.

(b) What does this mean about the graph of $y = x^2 + 4x + 4$?

A3 How can you tell when a quadratic function will only have one root?

A4 Calculate the roots of these quadratic functions and make a rough sketch of each on separate diagrams.

(a) $y = x^2 - 6x + 9$
(b) $y = -x^2 + 8x - 16$
(c) $y = x^2 + 12x + 36$
(d) $y = -x^2$
(e) $y = -x^2 - 4x - 4$
(f) $y = x^2 + 10x + 25$

A5 (a) If $y = x^2 + ax + 36$ has only one root, what is the value of a?

(b) If $y = -x^2 + 18x + b$ has only one root, what is the value of b?

A6 (a) What happens when you try to find the roots of $y = x^2 + 6$?

(b) What does this mean about the graph of $y = x^2 + 6$?

Here is one way to make a rough sketch of quadratic functions with no roots.

Example
Sketch the graph of $y = x^2 + 6x + 17$.

The roots are given by $x^2 + 6x + 17 = 0$. This does not factorise. So we will use the quadratic formula

$$x = \frac{-b \pm \sqrt{b^2 - 4ac}}{2a} = \frac{-6 \pm \sqrt{36 - 4 \times 1 \times 17}}{2} = \frac{-6 \pm \sqrt{-32}}{2}$$

Since we cannot find $\sqrt{-32}$ in the set of real numbers, the function has no roots. So the graph does not cut the x-axis at all. It is a U-shaped graph, but it could look like any of these:

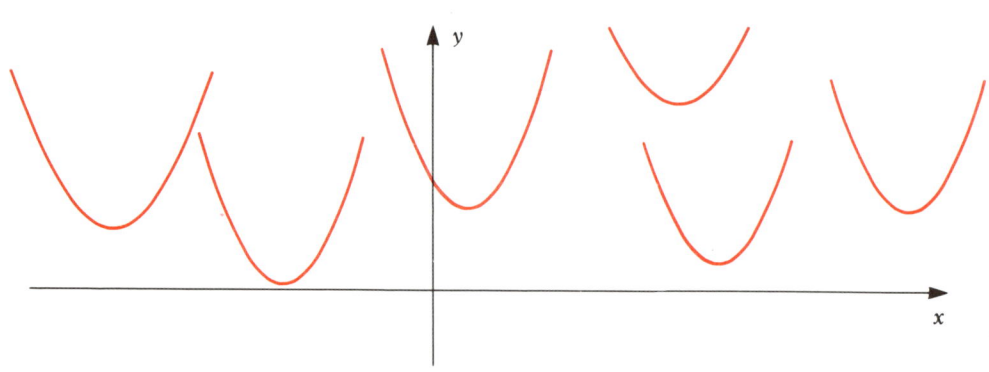

To get a better idea of the position of the graph, we 'complete the square' for the function.

$y = x^2 + 6x + 17$
$ = x^2 + 6x + 9 + 8$
$ = (x + 3)^2 + 8$

We need 9 to complete the square with the x^2 and the $6x$.

The 8 combined with the 9 still gives 17. So our equation has not changed.

This is of the form

$y = \text{(something)}^2 + \text{something else}$

The lowest value (something)^2 can have is 0. When that happens,

$y = \text{something else}$

So for our function the lowest value happens when $x + 3 = 0$, i.e. $x = -3$ and at that point $y = 8$.
This means the minimum turning point is $(-3, 8)$ and the sketch looks like this:

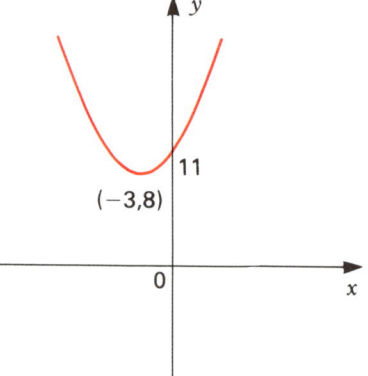

The graph cuts the y-axis when $x = 0$, so
$y = 0^2 + 6 \times 0 + 17 = 17$

A7 These quadratic functions have no roots. Use the method of 'completing the square' to find the coordinates of their turning points. Hence make a rough sketch of each on separate diagrams.

(a) $y = x^2 + 6x + 10$
(b) $y = x^2 - 4x + 6$
(c) $y = x^2 - 8x + 24$
(d) $y = x^2 + 10x + 30$
(e) $y = x^2 - 2x + 3$
(f) $y = x^2 + 5x + 7$

B Approximate roots of quadratic functions

All the roots of quadratic functions we have met so far have been exact. In reality this does not always happen.

Consider the quadratic function $y = x^2 - 6x + 2$. The roots are given by $x^2 - 6x + 2 = 0$.
This does not factorise, so we will try the quadratic formula

$$x = \frac{-b \pm \sqrt{b^2 - 4ac}}{2a}$$
$$= \frac{6 \pm \sqrt{36 - 4 \times 1 \times 2}}{2}$$
$$= \frac{6 \pm \sqrt{36 - 8}}{2}$$
$$= \frac{6 \pm \sqrt{28}}{2}$$

Since $\sqrt{28}$ is an irrational number, we cannot find it exactly. If the roots are required to 2 decimal places, then we take $\sqrt{28}$ to 3 decimal places, work through the formula with this value and round to 2 decimal places at the end.

So the roots are given by

$(+)$
$$x = \frac{6 + 5 \cdot 292}{2}$$
$$= \frac{11 \cdot 292}{2}$$
$$= 5 \cdot 646 = 5 \cdot 65 \text{ (to 2 d.p.)}$$

or

$(-)$
$$x = \frac{6 - 5 \cdot 292}{2}$$
$$= \frac{0 \cdot 708}{2}$$
$$= 0 \cdot 354 = 0 \cdot 35 \text{ (to 2 d.p.)}$$

B1 Calculate the roots of these quadratic functions to the required accuracy.

(a) $y = x^2 + 3x - 1$ (to 1 d.p.) (b) $y = 2x^2 + x - 4$ (to 2 d.p.)
(c) $y = 2x^2 + 7x + 2$ (to 2 d.p.) (d) $y = -5x^2 + 3x + 3$ (to 1 d.p.)
(e) $y = x^2 - x - 1$ (to 3 d.p.) (f) $y = -3x^2 - x + 1$ (to 2 d.p.)

C Iterative method to find approximate roots

It is easy enough to find approximate roots of quadratic functions by using the quadratic formula. If we have to find the approximate roots of a function other than a quadratic, however, this method is not available.

An iterative process is a method where successive applications result in a better approximation to the answer.

```
┌─────────────────┐    ┌─────────────────┐    ╱Is the    ╲
│ Start with an   │    │ Use the         │   ╱ approximation╲
│ approximation to│───▶│ approximation to│──▶╲  close     ╱──── Yes
│ the solution    │    │ give an even    │    ╲ enough? ╱
└─────────────────┘    │ better          │         │                │
         ▲             │ approximation   │         │ No          ┌──────┐
         │             └─────────────────┘                       │ STOP │
         │                                                       └──────┘
         └────────────────────────────────────────┘
```

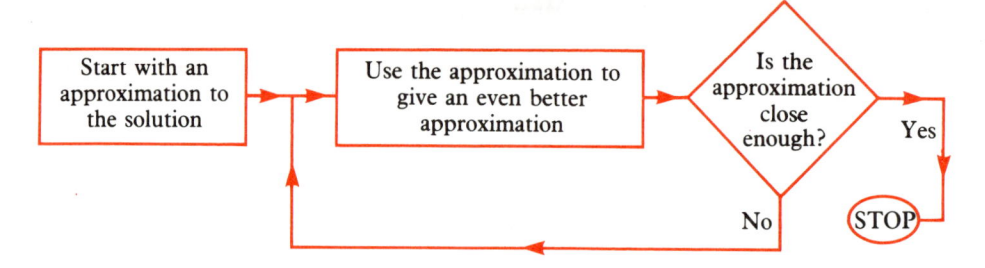

We used an iterative method for finding square roots in *Mathematics for Credit 1*.

Example
Find the root of $y = x^3 + 3x^2 - x - 5$ between 1 and 2, to 1 d.p.

When $x = 1$: $y = 1^3 + 3 \times 1^2 - 1 - 5 = 1 + 3 - 1 - 5 = -2$
This gives the point $(1, -2)$ on the curve.

When $x = 2$: $y = 2^3 + 3 \times 2^2 - 2 - 5 = 8 + 12 - 2 - 5 = 13$
This gives the point $(2, 13)$ on the curve.

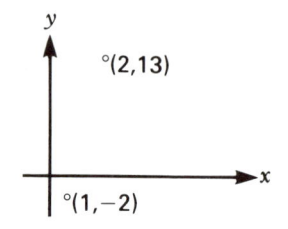

Since the function is a continuous one and goes from a point below the x-axis $(1, -2)$ to a point above the x-axis $(2, 13)$, it must cross the x-axis between 1 and 2.

So $\qquad 1 < \text{root} < 2$

We can use this to help get a better approximation to the root. Take the mid-point of 1 and 2.

When $x = 1 \cdot 5$: $y = 1 \cdot 5^2 + 3 \times 1 \cdot 5^2 - 1 \cdot 5 - 5 = 3 \cdot 375 + 6 \cdot 75 - 1 \cdot 5 - 5$
$\qquad\qquad\qquad\qquad\qquad\qquad\qquad\qquad = 3 \cdot 6 \text{ (to 1 d.p.)}$

This gives the point $(1 \cdot 5, 3 \cdot 6)$ on the curve

Since $(1 \cdot 5, 3 \cdot 6)$ is above the x-axis,

$$1 < \text{root} < 1 \cdot 5$$

The next value to take would be $x = 1 \cdot 25$ (the mid-point of 1 and 1·5) which gives $y = 0 \cdot 4$.

All this can be set out in a table like this

x-coordinate	y-coordinate		Root is between
	Positive (above x-axis)	Negative (below x-axis)	
1			–
2			1 and 2
1·5			1 and 1·5
1·25			1 and 1·25

We want the root correct to 1 decimal place so it seems sensible to try $x = 1·05$ or $1·15$ next since these will be crucial points when we round off.

x-coordinate	y-coordinate		Root is between
	Positive (above x-axis)	Negative (below x-axis)	
1			–
2			1 and 2
1·5			1 and 1·5
1·25			1 and 1·25
1·05			1·05 and 1·25
1·15			1·15 and 1·25

Hence the root is $x = 1·2$ (to 1 d.p.)

C1 Find the roots of the following functions to 1 decimal place in the given interval.

(a) $y = x^3 - 2x^2 + 3x - 5$ (between 1 and 2)
(b) $y = x^3 - 4x^2 + 3$ (between 3 and 4)
(c) $y = 2x^3 - x^2 - 3x + 1$ (between 0 and 1)
(d) $y = x^3 + x^2 - 7x + 6$ (between -4 and -3)
(e) $y = x^4 - 3x^3 + x^2 + x + 1$ (between 1 and 2)

C2 The function $y = 2x^3 + 15x^2 + 24x - 30$ has only one positive root. Find it correct to 1 decimal place.

D Solving linear and quadratic equations

Consider the quadratic function
$y = x^2 - x - 6$.

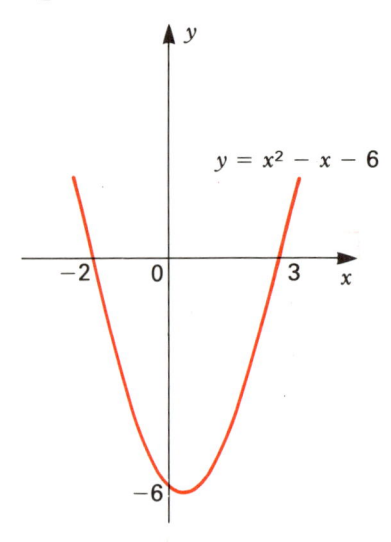

To see if a point, say (5, 14), lies on the curve, we must check whether its coordinates satisfy the equation of the curve.

Replacing x by 5 and y by 14 in $y = x^2 - x - 6$, we get

$14 = 5^2 - 5 - 6$
$14 = 25 - 5 - 6$

which is true.

Hence (5, 14) lies on the curve.

D1 For the curve whose equation is $y = x^2 + 7x - 2$, which of the following points lie on the curve?

(a) (3, 25)   (b) (−2, −12)   (c) (−7, 2)   (d) (4, 42)

D2 The equation of a circle centre (−3, 5) is
$x^2 + y^2 + 6x - 10y - 66 = 0$. Which of these points lie on the circle?

(a) (3, −1)   (b) (−11, 11)   (c) (5, −1)   (d) (2, 15)

D3 The equation of a hyperbola is
$x^2 + 6xy - 7y^2 - 8x + 8y - 20 = 0$. Which of the following points lie on the hyperbola?

(a) (−2, 0)   (b) (7, 1)   (c) (3, 4)   (d) (−5, 1)

D4 The point (2, p) lies on the curve whose equation is
$y = x^2 + 3x - 7$. By forming an equation in p, find the value of p.

D5 The point (q, 32) lies on the curve whose equation is
$y = 3x^2 + 3x - 4$. Form an equation in q and hence find the value of q.

D6 These points lie on the curve whose equation is
$x^2 + y^2 - 4x - 2y - 20 = 0$. By forming equations in the unknowns, find them.

(a) $(a, -2)$ (b) $(-1, b)$ (c) (c, c) (d) $(d, -d)$

The line $y = 10$ cuts the parabola whose equation is $y = x^2 - 8x + 17$ in two places.

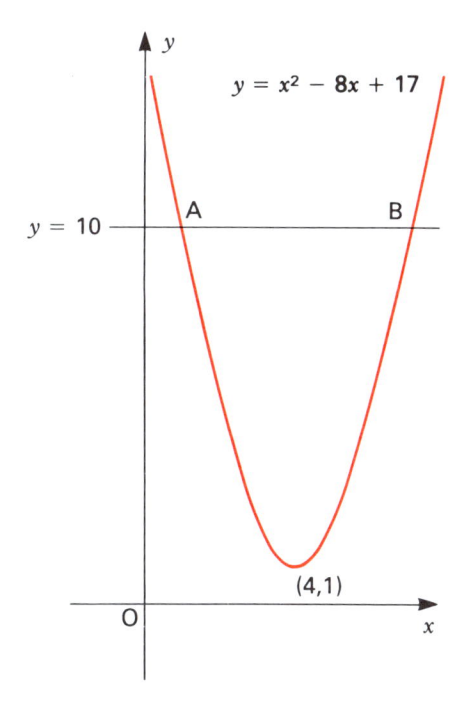

We can calculate the coordinates of the two points A and B using a similar method to the one used for points on curves.

Since A and B lie on the line $y = 10$, they both have y-coordinate 10, and so can be written as $(a, 10)$
If they lie on $y = x^2 - 8x + 17$, then
$$10 = a^2 - 8a + 17$$
$$a^2 - 8a + 7 = 0$$
$$(a - 1)(a - 7) = 0$$
$$a = 1 \text{ or } a = 7$$
The points are A (1, 10) and B (7, 10).

D7 (a) Calculate the coordinates of the points where the line $y = 20$ cuts the parabola $y = x^2 - 10x + 9$.

(b) Make a sketch to illustrate your answer.

D8 (a) Calculate the coordinates of the points where the line $x = 2$ cuts the parabola $y^2 = 8x$.

(b) Make a sketch to illustrate your answer.

D9 (a) The line $y = 9$ cuts the parabola $y = -x^2 - 4x + 5$ in one point. Find it.

(b) Make a sketch to illustrate your answer.

D10 For what values of k does the line $y = k$ **not** cut these parabolas?

(a) $y = x^2$ (b) $y = x^2 + 6x - 7$ (c) $y = x^2 - 3x - 4$

(d) $y = -x^2 - 4x - 6$

D11 (a) If a point lies on the line $y = x + 1$ and its x-coordinate is q, what is its y-coordinate?

(b) Use this to find the coordinates of the points of intersection of the line $y = x + 1$ and the parabola $y = x^2 - 3x - 4$.

When finding the points of intersection of a line and a parabola we are solving the equation of a line (a **linear** equation) and the equation of the parabola (a **quadratic** equation) simultaneously.

Example
Find the coordinates of the points of intersection of the line $y = x - 9$ and the parabola $y = -x^2 - x + 6$.

The coordinates of any point on the line must satisfy the equation $y = x - 9$. The coordinates must also satisfy $y = -x^2 - x + 6$ since the point also lies on the parabola. Now these x- and y-coordinates are for the same point so they must also be equal.

Hence we have two equations with the same unknowns in each:

$y = x - 9$ and $y = -x^2 - x + 6$

Therefore $x - 9 = -x^2 - x + 6$

This is a quadratic equation and can be solved as before.

$x - 9 + x^2 + x - 6 = 0$
$x^2 + 2x - 15 = 0$
$(x + 5)(x - 3) = 0$
$x = -5$ or $x = 3$

Since there are two solutions, the line cuts the parabola in two points.

To find the y-coordinates of the two points, we substitute the x-coordinates into one of the original equations. The equation of the line is easier to use.

Substituting $x = -5$ and $x = 3$ into $y = x - 9$, we get $y = -14$ and $y = -6$.
The points of intersection are $(-5, -14)$ and $(3, -6)$

It is always good practice to make a sketch as it gives an indication as to whether the points obtained are feasible.

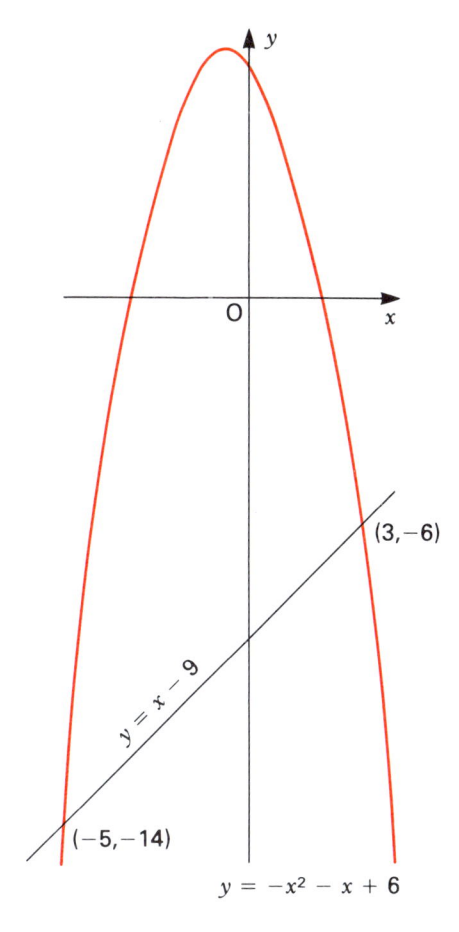

(Sketch: parabola $y = -x^2 - x + 6$ with line $y = x - 9$ intersecting at $(3, -6)$ and $(-5, -14)$.)

D12 Find the coordinates of the points of intersection of the following lines and parabolas. Illustrate each answer with a sketch.

(a) $y = x - 1$, $y = x^2 - 6x + 5$

(b) $y = -x + 10$, $y = x^2 + 2x - 8$

(c) $y = 2x + 7$, $y = -x^2 - 6x + 7$

(d) $y = 4x - 5$, $y = 2x^2 - 8x + 11$

(e) $2x - y = 13$, $y = -x^2 + 6x - 8$

(f) $2y - x = 6$, $y = \frac{1}{2}x^2 + 2x + 4$

(g) $x + y = 8$, $y = -2x^2 + 3x + 14$

(h) $2x + y = 5$, $y = 4x^2 - 12x - 1$

D13 Find the coordinates of the points of intersection of these lines and circles. Use the same method as for parabolas and comment on the results.

(a) $x + 2y = 12$, $x^2 + y^2 - 6x - 4y + 8 = 0$

(b) $2x - y = -6$, $x^2 + y^2 - 8x - 6y + 21 = 0$

Example
Find the coordinates of the points of intersection of the line
$4x + 3y = 12$ and the circle $x^2 + y^2 - 12x + 8y + 27 = 0$.

First we must rearrange the linear equation into a form we can substitute into the equation of the circle.

$4x + 3y = 12$
$4x = 12 - 3y$
$x = 3 - \tfrac{3}{4}y$

Replacing x by $3 - \tfrac{3}{4}y$, the equation of the circle becomes

$(3 - \tfrac{3}{4}y)^2 + y^2 - 12(3 - \tfrac{3}{4}y) + 8y + 27 = 0$
$9 - \tfrac{9}{2}y + \tfrac{9}{16}y^2 + y^2 - 36 + 9y + 8y + 27 = 0$

$144 - 72y + 9y^2 + 16y^2 - 576 + 144y + 128y + 432 = 0$

Multiply by 16 to get rid of the fractions.

$25y^2 + 200y = 0$ ← *Collecting like terms.*
$y^2 + 8y = 0$
$y(y + 8) = 0$
$y = 0$ or $y = -8$

Substituting into the equation of the line we get

$x = 3 - \tfrac{3}{4} \times 0$ $x = 3 - \tfrac{3}{4} \times (-8)$
$= 3 - 0$ $= 3 + 6$
$x = 3$ $x = 9$

So the coordinates of the points of intersection are $(3, 0)$ and $(9, -8)$.

> **D14** Solve these equations by the method of substitution, as in the previous example.
>
> (a) $x + y = 7$, $x^2 + y^2 = 49$
>
> (b) $2x + y = 10$, $x^2 + y^2 - 10y = 0$
>
> (c) $y = 1 - 2x$, $x^2 + xy + y^2 = 7$
>
> (d) $2x + 3y = 31$, $x^2 + y^2 + 4x - 6y - 156 = 0$
>
> (e) $2x - 3y = 5$, $x^2 + y^2 + 4x - 4y - 5 = 0$
>
> (f) $2x - 2y = 1$, $4x^2 + 4y^2 - 8x - 8y - 33 = 0$
>
> (g) $2x - 8y = -25$, $2x^2 + 2y^2 + 16x + 15 = 0$
>
> (h) $9x - 3y = 46$, $9x^2 + 9y^2 - 54x - 36y - 8 = 0$

E Finding the quadratic function from its graph

E1 (a) Find the roots of $y = x^2 + x - 12$ and $y = 2x^2 + 2x - 24$.

(b) Draw a rough sketch of their graphs on the same diagram.

E2 (a) Find the roots of $y = -x^2 + 6x - 5$ and $y = -3x^2 + 18x - 15$.

(b) Draw a rough sketch of their graphs on the same diagram.

E3 (a) Find the roots of $y = x^2 + 9x + 18$ and make a rough sketch of its graph.

(b) How could you draw the rough sketch of $y = kx^2 + 9kx + 18k$ on the same diagram? (Assume $k > 0$.)

This illustrates an important property:

If two quadratic functions have the same roots, then the equation of one is a multiple of the other.

Example
Find the equation of the quadratic function whose graph is

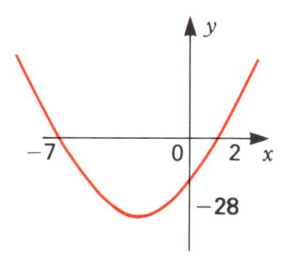

Remember how we calculated the roots of a quadratic function, for example

$y = x^2 - 2x - 24$

The roots are given by $x^2 - 2x - 24 = 0$
$(x - 6)(x + 4) = 0$
Either $x - 6 = 0$ or $x + 4 = 0$
$x = 0$ or $x = -4$

To help find the equation of our quadratic function we will do this **backwards**.

So the roots are $x = -7$ or $x = 2$

Working backwards we have
$x + 7 = 0$ or $x - 2 = 0$
$(x + 7)(x - 2) = 0$
$x^2 + 5x - 14 = 0$

But since we now know that there is more than one quadratic function with roots of -7 and 2, we can use the result about any of these functions being a multiple of $x^2 + 5x - 14$. The equation we are looking for will be of the form

$y = k(x^2 + 5x - 14)$
$y = kx^2 + 5kx - 14k$

To find k we use the other point we are given on the curve.
Substituting $(0, -28)$ into this equation

$$-28 = 0 + 0 - 14k$$
$$-28 = -14k$$
$$k = 2$$

So the equation of the curve is
$$y = 2(x^2 + 5x - 14) = 2x^2 + 10x - 28$$

E4 Find the equations of the quadratic functions from their rough sketches.

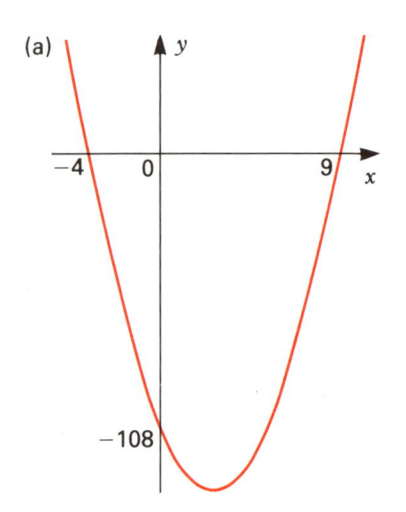

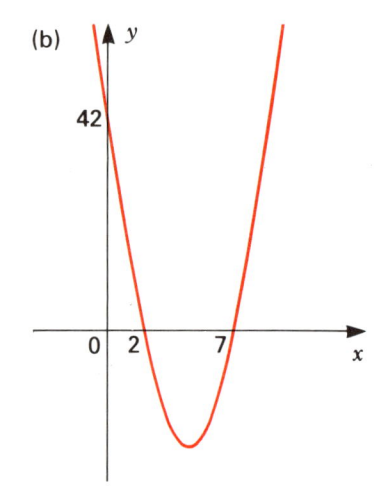

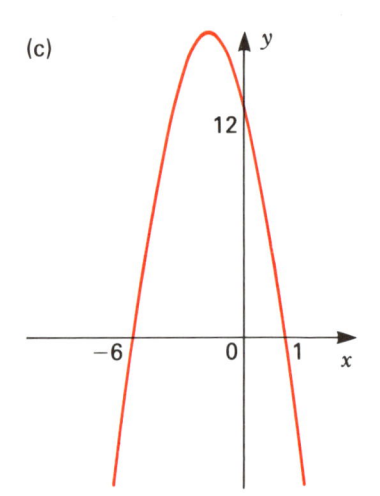

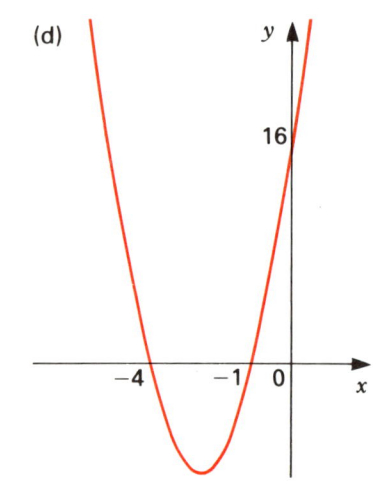

(a) Parabola opening upwards, crossing x-axis at -4 and 9, y-intercept at -108.

(b) Parabola opening upwards, crossing x-axis at 2 and 7, y-intercept at 42.

(c) Parabola opening downwards, crossing x-axis at -6 and 1, y-intercept at 12.

(d) Parabola opening upwards, crossing x-axis at -4 and -1, y-intercept at 16.

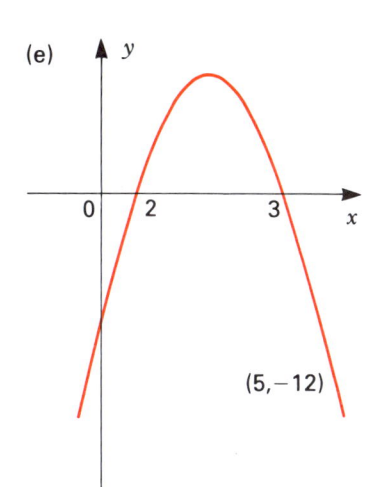

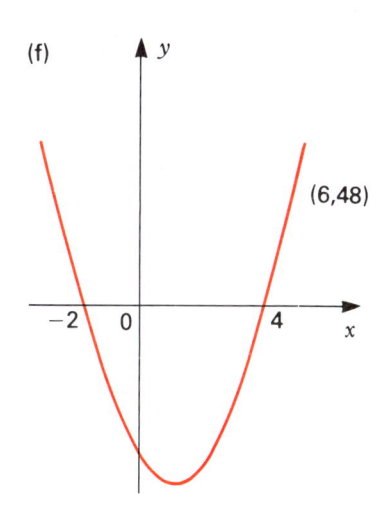

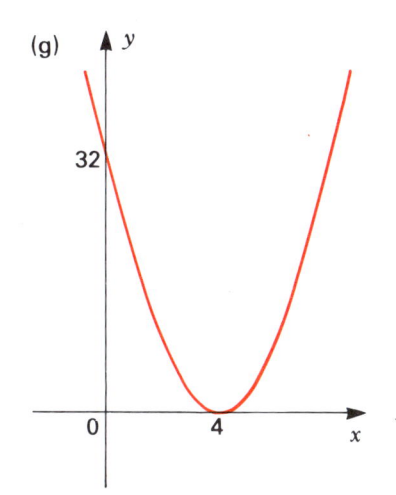

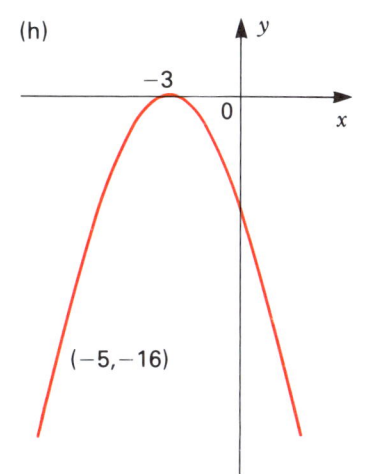

(e) [Graph: parabola opening downward through x=2 and x=3, with point (5,−12)]

(f) [Graph: parabola opening upward through x=−2 and x=4, with point (6,48)]

(g) [Graph: parabola opening upward with vertex at x=4, y-intercept 32]

(h) [Graph: parabola opening downward through x=−3, with point (−5,−16)]

F Quadratic inequations

F1 (a) Copy and complete the table for the quadratic function
$y = x^2 - 3x - 4$.

x	−3	−2	−1	0	1	2	3	4	5	6
$x^2 - 3x - 4$										

(b) Hence draw its graph on 2 mm squared paper using suitable scales.

(c) Beside each of the 10 points on the graph, write the value of $x^2 - 3x - 4$ (the y-coordinate).

(d) Where are the points on the graph where $x^2 - 3x - 4 > 0$? Give the range of values for x for these points.

159

(e) Where are the points on the graph where $x^2 - 3x - 4 < 0$? Give the range of values for x for these points.

F2 (a) Draw the graph on 2 mm squared paper for the function $y = -x^2 - 2x + 3$ for $-6 \leqslant x \leqslant 4$.

(b) Give the range of values for x for points on the graph where $-x^2 - 2x + 3 > 0$.

(c) Give the range of values for x for points on the graph where $-x^2 - 2x + 3 < 0$.

F3 (a) Draw the graph on 2 mm squared paper for the function $y = x^2 - 7x + 10$ for $-1 \leqslant x \leqslant 7$.

(b) Give the range of values for x for points on the graph where $x^2 - 7x + 10 \geqslant 0$.

F4 (a) If we want to find points where $x^2 + 4x - 5 > 0$, what graph should be drawn?

(b) Make a rough sketch of this graph.

(c) Where are the points on the graph where $x^2 + 4x - 5 > 0$?

(d) Using the rough sketch, give the range of values for x for these points.

Inequations like $x^2 - 3x - 4 > 0$, $-x^2 - 2x + 3 \leqslant 0$ and $x^2 - 7x + 10 < 0$ are called **quadratic inequations** because one side is a quadratic function.

They can be solved by following these steps:

> (1) Make a rough sketch of the function.
>
> (2) Identify the part of the curve the inequation is giving (> 0 is **above** the x-axis; < 0 is **below** the x-axis).
>
> (3) Use the rough sketch to find the values of x for that part of the curve.

Example
Solve $x^2 + 9x + 14 \geqslant 0$.

Make a rough sketch of
$y = x^2 + 9x + 14$.
The roots are given by
$x^2 + 9x + 14 = 0$
$(x + 2)(x + 7) = 0$
$x = -2$ or $x = -7$

The curve cuts the y-axis at 14. So the sketch looks like this.

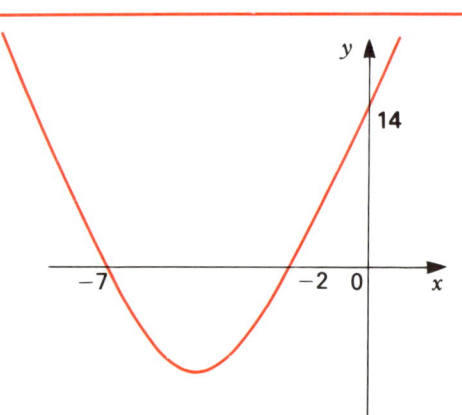

Now we want $x^2 + 9x + 14 \geq 0$

This is the part of the curve above or on the x-axis.

Two parts of the curve are above the x-axis:

$$x < -7 \text{ or } x > -2$$

Since points can also lie **on** the x-axis, this gives

$$x \leq -7 \text{ or } x \geq -2$$

F5 Solve the following quadratic inequations.

(a) $x^2 - x - 6 > 0$

(b) $x^2 - 2x - 15 > 0$

(c) $2x^2 - 12x + 18 \geq 0$

(d) $9 - x^2 < 0$

(e) $x^2 + 16 < 0$

(f) $x^2 - x - 42 < 0$

(g) $-x^2 - 10x - 25 \geq 0$

(h) $-5x^2 - 20 < 0$

(i) $2x^2 - 22x + 48 < 0$

(j) $3x^2 + 6x - 45 \leq 0$

Consolidation 3

A The sine rule and the area of a triangle

A1 Calculate the sides and angles marked **?**.

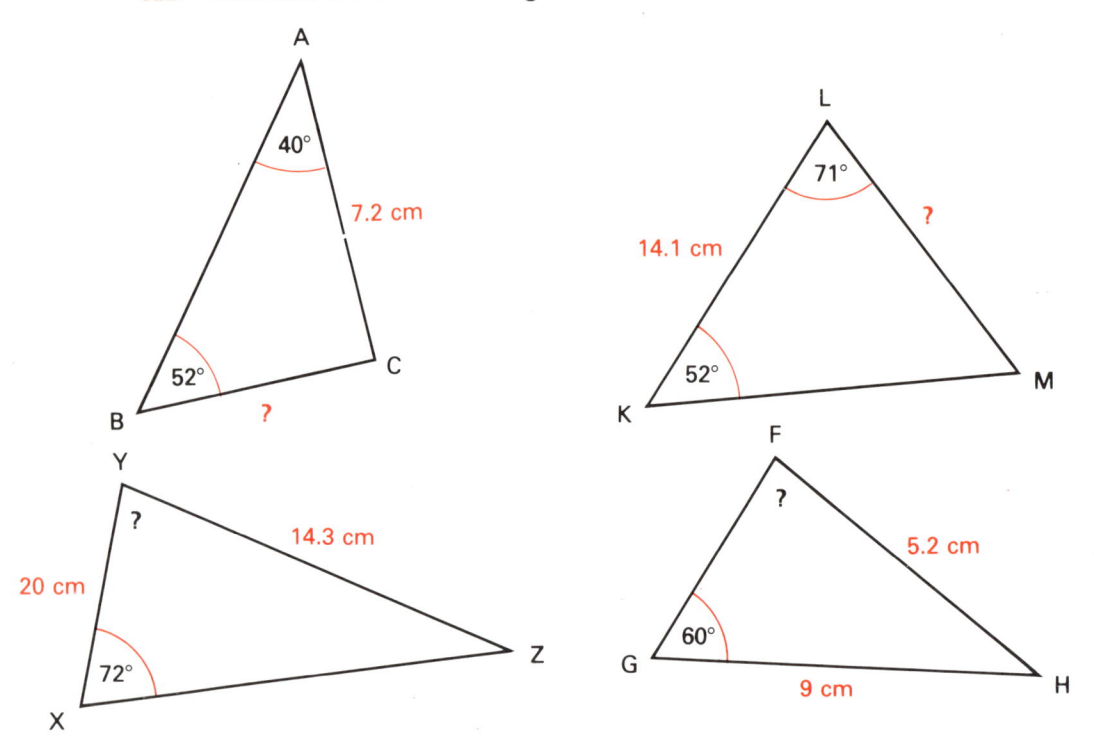

A2 A coastguard observes a boat sailing due east 2 miles away on a bearing of 040°. An hour later the bearing of the boat is 073°. Calculate the speed of the boat.

A3 Two construction workers standing on a building vertically above each other see the teaboy waiting on the ground with their tea.

If the workers are 20 m apart and the angles of depression of the teaboy are 39° and 51°, calculate the height of the workers above the ground.

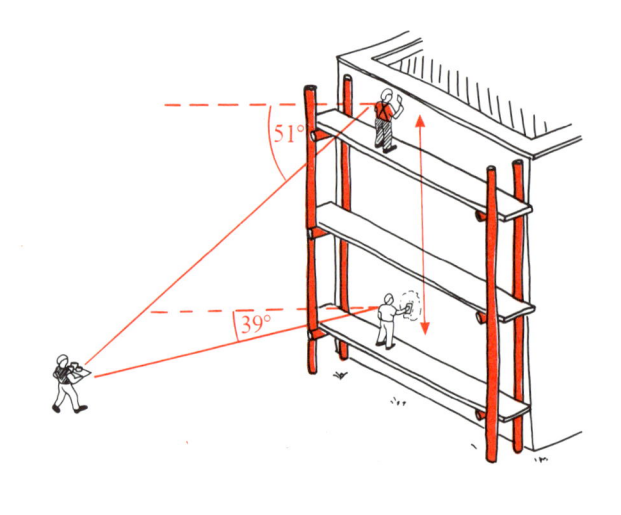

A4 Calculate the area of each of these triangles.

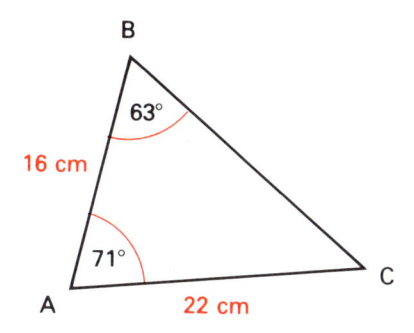

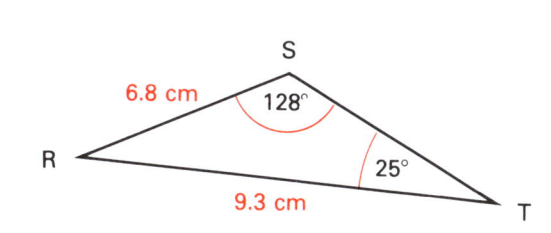

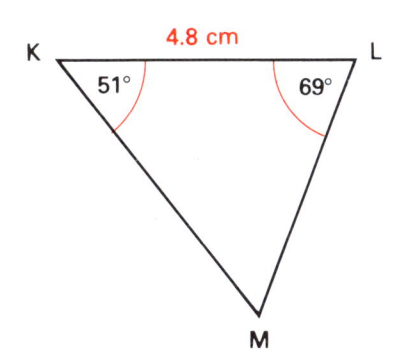

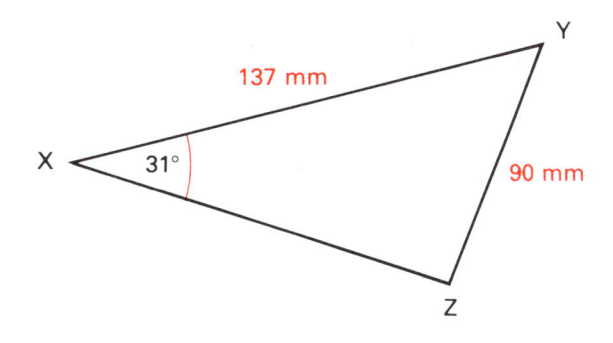

A5 Here is the side view of a set of five-a-side football goals. Calculate the area of net required for this side.

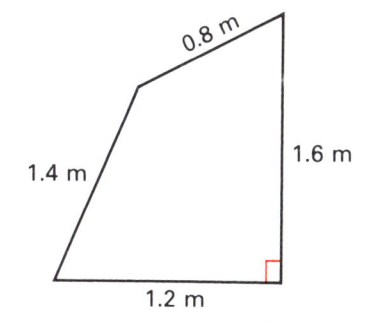

A6 Find the possible values of the angle marked **?**.

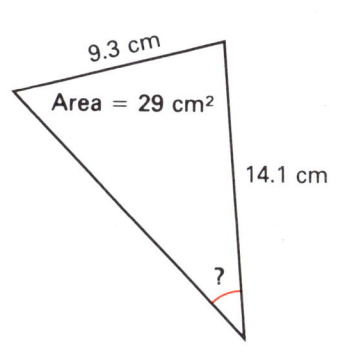

B Quadratic equations

B1 Solve the following equations where possible. Where answers are not exact, give solutions correct to 1 decimal place.

(a) $10x^2 + x - 2 = 0$

(b) $2x^2 - 3x = 14$

(c) $3(x + 1)^2 = x + 5$

(d) $1 - 2(x - 2) = 2(2x + 1)^2$

(e) $(5x + 1)(3x + 2) = (x - 4)^2$

(f) $(x + 3)(2x - 1) = 4 - (x - 1)^2$

(g) $3(x - 2)(x + 1) = -(x - 2)^2$

(h) $2x(3x + 1) + (x + 3)^2 = (x + 2)(x - 5)$

(i) $(3x + 1)(2x + 3) - 11x = (5x - 1)(x + 1)$

(j) $-(x - 3)^2 + (2x - 1)(x + 2) = 3x(x - 2)$

B2 Solve these equations where possible. Where answers are not exact, give solutions correct to 1 decimal place.

(a) $\dfrac{4}{x} - \dfrac{10}{x - 1} = 1$

(b) $\dfrac{3}{2x} + \dfrac{5}{x + 2} = 1$

(c) $\dfrac{1}{x + 7} - 1 = \dfrac{3}{x - 1}$

(d) $10(x - 1) + \dfrac{10}{x - 1} = 29$

(e) $\dfrac{3}{4} = \dfrac{3}{5x - 17} - \dfrac{2}{5x - 12}$

(f) $\dfrac{x - 1}{x - 2} - \dfrac{5}{6} = \dfrac{2 - x}{x - 1}$

(g) $\dfrac{7}{3} + \dfrac{x - 6}{x + 3} = \dfrac{2x - 1}{x}$

(h) $\dfrac{1}{x} + \dfrac{2}{x - 1} + \dfrac{3}{x - 1} = 0$

(i) $\dfrac{2x + 7}{x + 1} = \dfrac{8}{x + 4} + \dfrac{2(x + 5)}{x + 2}$

(j) $\dfrac{5 + x}{1 + x} = \dfrac{2 + 2x}{3 + x} + \dfrac{1 - x}{4 + x}$

B3 One side of a rectangular sheet of paper is 4 cm longer than the other. When a strip of width 2 cm is cut off all round, the remaining area is 32 cm². Find the dimensions of the original sheet of paper.

B4 A ball is thrown upwards so that its height above the ground after t seconds is $(30t - 5t^2)$ m.

(a) Find when the ball is 20 m above the ground. Explain your answer.

(b) When does the ball hit the ground?

B5 A toy is bought for £x. It is later sold at £26·84 giving a profit of x%. Calculate x.

B6 The sum of a number and 12 times its reciprocal is 8. Find the number.

B7 This square and rectangle both have the same perimeter.

6 cm

The rectangle is 6 cm broad and its area is 16 cm² less than the square. Find the area of the square.

C Time for a holiday!

C1 Two petrol stations have different ways of advertising the price of their petrol. Exacto petrol is priced at 39·6p per litre whereas one gallon of Scottich Petroleum's petrol is £1·87.

(a) Which petrol is the more expensive?

(b) How much more expensive is it as a percentage rounded to the nearest whole number?

1 gallon ≈ 4·546 litres

C2 Petrol is £1·87 per gallon in Britain. If it is 5·20 francs per litre in France, calculate how much more expensive petrol is in France compared to Britain. Give your answer as a percentage rounded to the nearest whole number.
(Suppose £1 = 10·20 francs.)

C3 Carol changed £900 into dollars at a rate of $1·54 to the £. She spent $1100 and wanted to have £180 left when she returned to Scotland.

What is the highest exchange rate that Carol needs to find to achieve this?

C4 Sheila and Drew changed £700 into currency for their holiday to Portugal. The rate of exchange was 240 escudos to the £.

During their holiday they spent 131 000 escudos and changed the remainder back into sterling on their return to Scotland.

Calculate how much money, in sterling, they lost if the rate for exchanging escudos to sterling was 248 escudos to the £.

C5 Mr and Mrs Campbell and their two sons aged 10 and 14 are going on holiday with their caravan to Holland. Their car is 4·37 m long and their caravan is 3·61 m long.

The distance from the Campbell's home to Harwich is 450 miles. Mr Campbell estimates that they will travel 700 miles while they are in Holland.

Petrol costs £1·96 per gallon in Britain and 1·54 guilders per litre in Holland.

(a) If the Campbell's car does 25 miles per gallon on average, calculate the cost of their travelling arrangements.

HARWICH HOOK OF HOLLAND / HOOK OF HOLLAND HARWICH			
STANDARD SINGLE FARES			
PASSENGERS:	ADULTS		£13·00
	CHILDREN (4 and under 14)		£ 7·00
	Children under 4 Free		
VEHICLES:	Cars, Motor caravans, Minibuses, Vans and Motorcycle combinations		
	OVERALL LENGTH not exceeding	4·00 m	£53·00
		4·50 m	£63·00
		5·50 m	£66·00
	Over 5·50 m per additional metre or part thereof		£11·00
TOWED CARAVANS/TRAILERS:			
	OVERALL LENGTH not exceeding	3·00 m	£10·00
		6·00 m	£18·00
	Over 6·00 m per additional metre or part thereof		£12·00

(£1 = 3·38 guilders; 1 gallon = 4·546 litres)

(b) How could the family minimise petrol costs?

D Linear programming

D1 Archie is making a path from his house to his garage at the end of his garden. It has to be at least 36 m long to reach the garage.

He is going to use two types of slab. They are both the same width and are either 1 m long or 30 cm long and cost £4 and £2 each respectively.

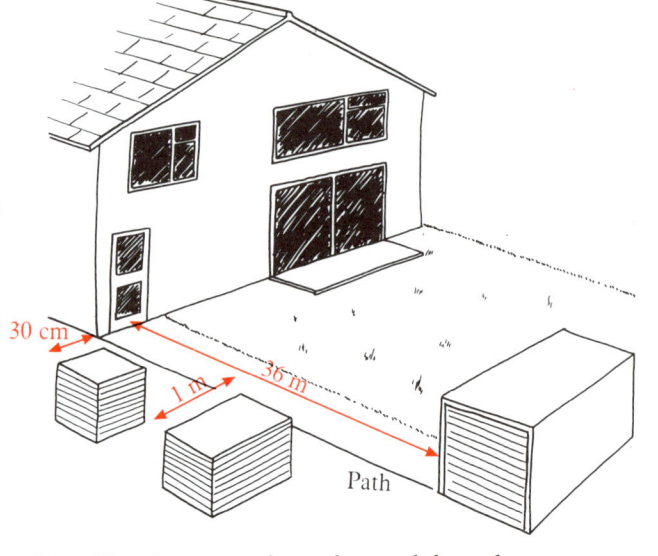

Archie estimates it takes him 40 minutes to lay a large slab and 10 minutes to lay a small one.

If Archie has £180 to spend on slabs, how many of each kind should he buy if he wants to lay his path as quickly as possible? How long will it take him?

D2 Ann is a picture-framer and produces two kinds of pictures. 'Squares' are square pictures and need 100 cm of framing. 'Oblongs' are rectangular pictures and need 150 cm of framing. Both pictures need 500 cm² of glass. Ann has 9000 cm of framing and 35 000 cm² of glass available.

(a) If she makes a profit of £3 on a 'square' and £4 on an 'oblong', how many of each kind should she make for maximum profit? What is that profit?

(b) If she changes her prices so that she now makes £3 on a 'square' and £6 on an 'oblong', how many of each should she now make for maximum profit? What is that profit?

(c) In each case (a) and (b), work out how much framing and how much glass Ann would have left over.

E Graphs 3

E1 Make rough sketches of the following parabolas, marking the turning points and the points of intersection with the axes where appropriate.

(a) $y = x^2 + 3x - 10$ (b) $y = -x^2 + 5x - 4$

(c) $y = x^2 - 4x + 9$ (d) $y = -x^2 - 2x - 1$

(e) $y = 2x^2 + 7x + 3$ (f) $y = 2x^2 - 12x + 18$

E2 Find the roots of the following functions to 1 decimal place in the given interval.

(a) $y = x^3 - 8x + 7$ (between 2 and 3)

(b) $y = 2x^3 - x^2 - 6x + 2$ (between 1 and 2)

E3 Solve these equations where possible.

(a) $x + y = 7$, $y = x^2 + 8x + 7$

(b) $7x + y = 40$, $x^2 + y^2 - 4x - 2y - 20 = 0$

(c) $3x - y + 6 = 0$, $x^2 + y^2 - 4x - 12 = 0$

(d) $3x - 2y = 8$, $x^2 + y^2 + 6x + 4y = 0$

(e) $2x + 8y = 61$, $2x^2 + 2y^2 - 8x - 20y + 41 = 0$

(f) $4x + 2y = -3$, $2x^2 + 2y^2 - 4y - 3 = 0$

E4 Find the equations of these quadratic functions from their rough sketches.

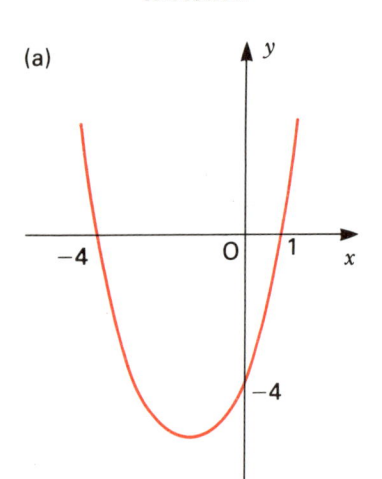

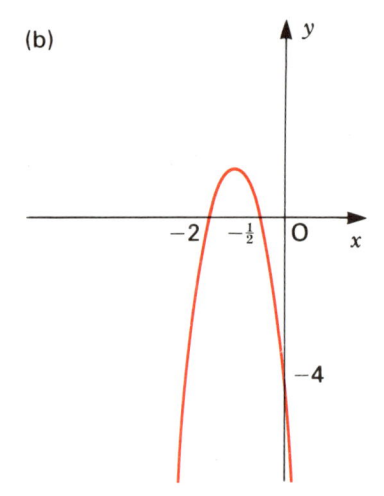

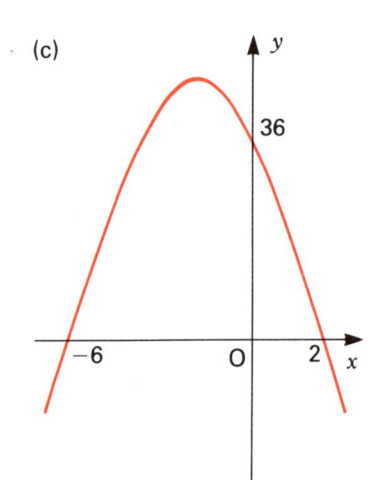

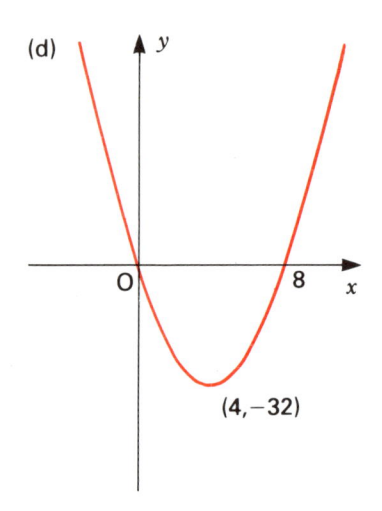

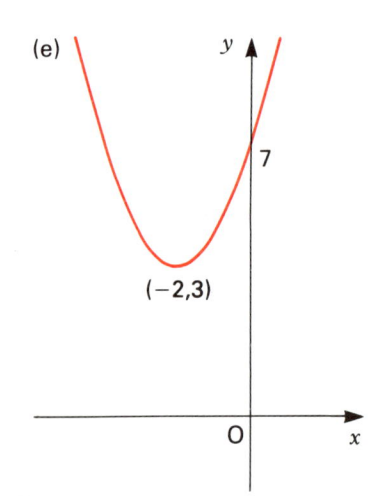

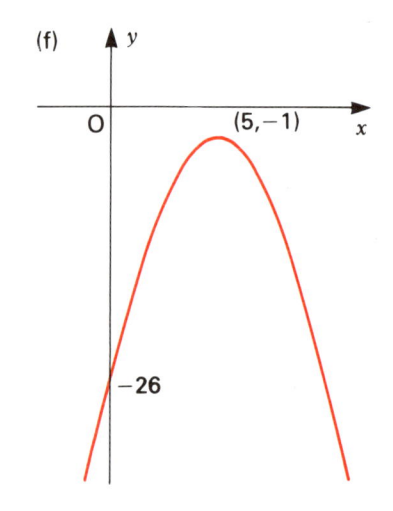

(a) Graph with x-intercepts at -4 and 1, y-intercept at -4.

(b) Graph with x-intercepts at -2 and $-\tfrac{1}{2}$, y-intercept at -4.

(c) Graph with x-intercepts at -6 and 2, y-intercept at 36.

(d) Graph with x-intercepts at 0 and 8, vertex at $(4, -32)$.

(e) Graph with vertex at $(-2, 3)$, y-intercept at 7.

(f) Graph with vertex at $(5, -1)$, y-intercept at -26.

168

E5 Solve these.

(a) $x^2 + 2x - 15 \leq 0$

(b) $x^2 + x - 2 > 0$

(c) $-x^2 - 4x - 3 \geq 0$

(d) $x^2 + 2x + 2 > 0$

(e) $-x^2 + 11x - 28 < 0$

(f) $x^2 - 9x \geq 0$

General review

A Book Y1 review

A1 Find the missing numbers in the following.

(a) ? → × 2·5 → 18·5   (b) ? → × 0·3 → 0·42
(c) ? → ÷ 1·6 → 3·2   (d) ? → ÷ 0·9 → 24·7

A2 If $a = 5$, $b = 2$ and $c = 10$, find the values of

(a) $ab + bc$ (b) $\left(\dfrac{a}{b}\right)^2 - c$ (c) $\dfrac{a + 2b}{b + 3c}$

A3 One packet of crisps sells for 16p. They are bought by a shopkeeper in boxes of 48 for £6.

(a) Write down an expression for the selling price of x packets of crisps.

(b) Write down an expression for how much the shopkeeper pays for x packets of crisps.

(c) Write down an expression for the profit made by the shopkeeper on x packets of crisps.

A4 Claire, Nicholas and Vicki share a £24 000 pools win. Nicholas gets twice as much as Claire, and Vicki gets 1·5 times as much as Nicholas.

How much do they each receive?

A5 Work out the following.

(a) $-7 - 5$ (b) $-3 - (-2)$ (c) $13 - (-12)$
(d) -5×-3 (e) $-6 \div 3$ (f) $(-4)^2$

A6 Solve these equations

(a) $2x - 4 = 6$ (b) $3x - 6\cdot3 = -9\cdot6$
(c) $(x + 2)(x - 1) = x^2 + 4$ (d) $(x + 4)(x - 3) = x(x - 5)$

A7 The length of the equator is roughly 40 506 km. Calculate the corresponding diameter of the Earth.

A8 Multiply out these brackets.

(a) $4(x - 2y)$ (b) $(x + 5)(x - 3)$ (c) $(p + 2)(p - 2q + 7)$
(d) $(1 + 2b)^2$ (e) $(9 - 3y)^2$

A9 William's father is 50 years older than he is. In 4 years, William will be half as old as his father. Work out their ages.

A10 Donna earns £14 500. Her salary is to be increased by 6%.

(a) Calculate her new salary.

(b) What would you multiply $14 500 by to get this new amount?

A11 These road signs show the gradients of hills. Rearrange them in order of steepness – steepest first.

1:6   10%   1:5

A12 (a) Use a diagram like this to explain clearly why

$(a + b)^2 = a^2 + 2ab + b^2$

(b) Use a similar diagram to prove the expansion of $(2x + 3)(x + 4)$.

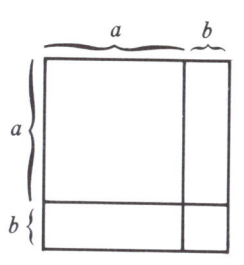

B Book Y2 review

B1 Which of the following tables shows a linear relationship?

(a)
a	2	3	6	8	10
b	5	7	10	14	22

(b)
p	1	2	3	4	5
q	1	4	9	16	25

(c)

x	10	20	30	40	50
y	31	61	91	121	151

B2 (a) Copy and complete the table below, which shows how the angle at the centre changes when a polygon is drawn inside a circle.

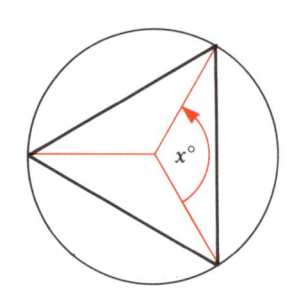

Number of sides (n)	3	4	5	6	7	8	9
Angle at centre ($x°$)	120	90					

(b) Draw the graph of (n, x) and continue it.

(c) Use the graph to find
  (i) the angle at the centre of a decagon
  (ii) the number of sides that a shape with an angle at the centre 24° has

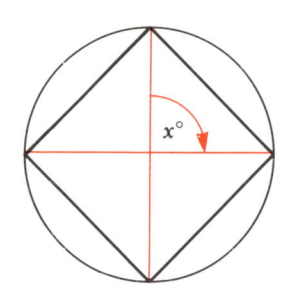

B3 The number of spectators at a football match was estimated to be 32 500 to the nearest hundred. Write this as an interval approximation.

B4 Jenny is standing 25 m away from the bottom of the church tower. She looks up at the top at an angle of elevation of 52°.

Calculate the height of the tower.

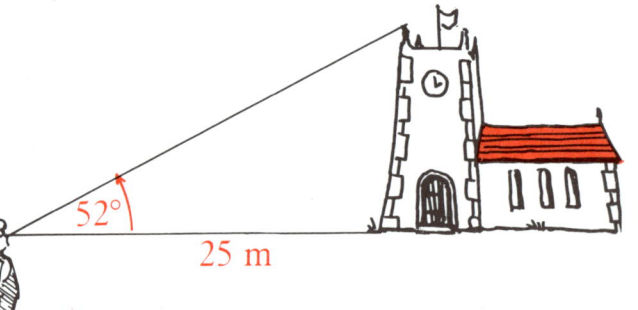

B5 Gordon mixes 5 litres of white paint with 2 litres of yellow paint to give a lemon shade.

Work out how many ml of white paint are in every litre of lemon.

B6 Factorise these.

(a) $5ab - 10b^2$ (b) $xy^2 - x^2y$

B7 Calculate the angles of this kite, and the length marked x cm.

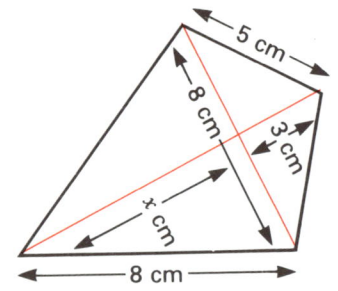

B8 The mean weight of a school's five-a-side football team is 45 kg.

They use two extra players to compete in a local seven-a-side tournament. The mean weight of the team goes up to 47 kg.

What is the mean weight of the additional players?

B9 The cooking time of a chicken can be worked out using the rule

$t = 25w + 20$

where t is the time the chicken takes to cook and w is its weight in pounds.

(a) Copy and complete this table.

w	2	3	4	5	6	7
t						

(b) Draw the graph of (w, t).

(c) Is t proportional to w? Explain your answer.

B10 Calculate the areas of the following shapes.

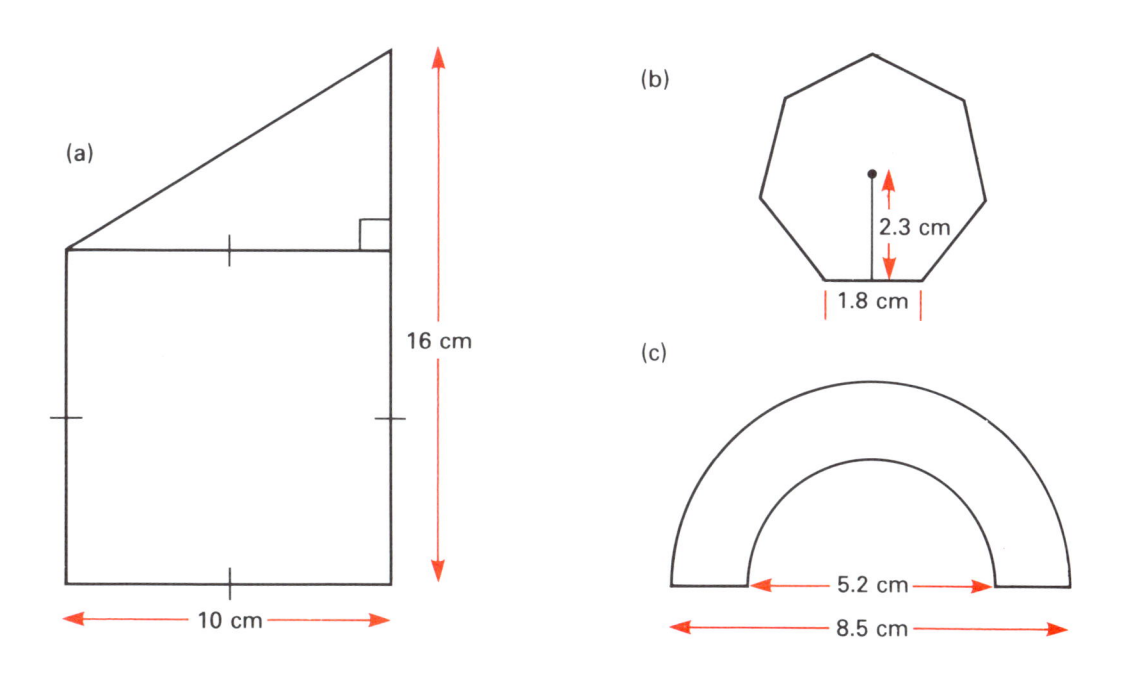

(a)

(b)

16 cm

10 cm

2.3 cm

1.8 cm

(c)

5.2 cm

8.5 cm

B11 A builder's merchant sells cement in two sizes of bag – 25 kg and 20 kg. Their lorry can carry a maximum of 500 kg at a time.

(a) Write down an expression for the total weight of cement in x large bags and y small bags.

(b) Write down an equation which says that the total weight of an order is 500 kg.

(c) Draw the graph of the equation you have stated. Shade the part of the graph which shows the weights which the lorry cannot carry.

C Book Y3 review

C1 Without drawing graphs, write down the gradient, and the intercept on the y-axis, of these lines.

(a) $y = -2x + 7$ (b) $x + y = 7$ (c) $3x - 2y + 4 = 0$

C2 Cameron scores $\frac{18}{25}$ in a History text and $\frac{35}{60}$ for Maths. By comparing percentages, work out which of these is a better mark.

C3 Decide which of the following shows direct and which shows inverse proportionality. Give reasons for your answers.

(a) The cost of a bus trip depends on the number of people who go.

(b) The cost of ribbon depends on the length bought.

(c) Henry VIII had 6 wives. How many wives did Henry VII have?

C4 Here is a list of the average monthly summer temperatures in Majorca and Yugoslavia.

	Apr.	May	Jun.	Jul.	Aug.	Sept.	Oct.
Majorca (°F)	66	71	78	84	83	80	72
Yugoslavia (°F)	61	70	76	81	80	75	65

(a) What is the median temperature for each place?

(b) What is the mean temperature for each place?

(c) What is the range of temperature for each place?

(d) Write a short report comparing the two sets of temperatures.

C5 Neil has a bag of sweets. He gives his brother Ian 40% of the sweets and his friend Donna 30% of what is left.

What percentage of the sweets did

(a) Donna get?   (b) Neil get?

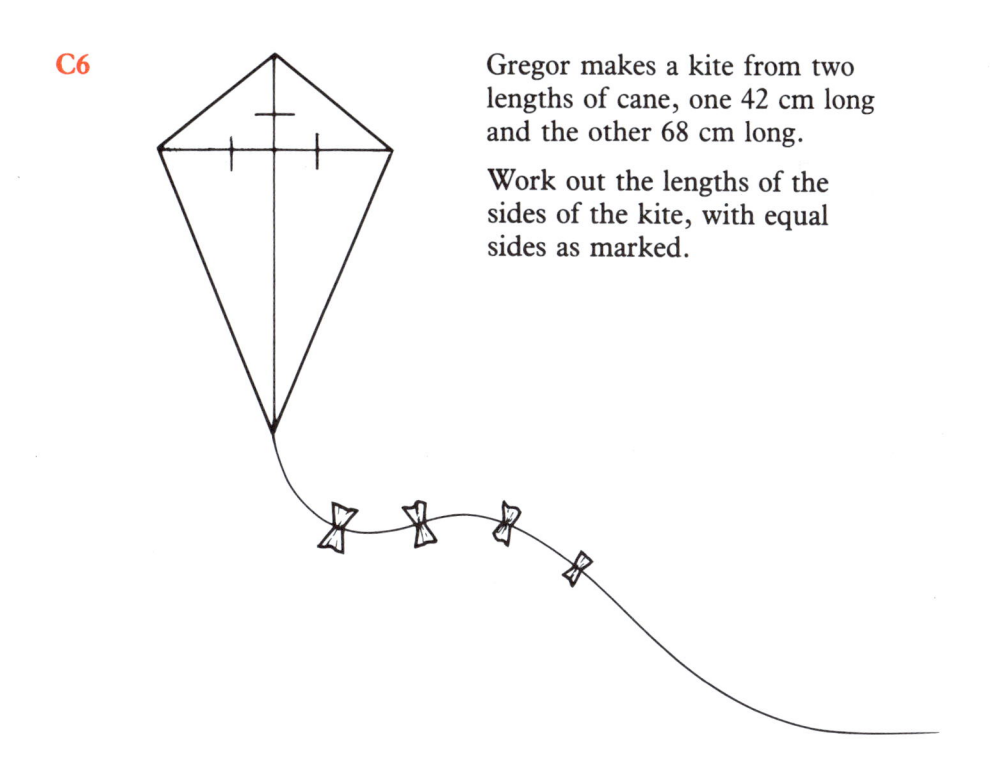

C6 Gregor makes a kite from two lengths of cane, one 42 cm long and the other 68 cm long.

Work out the lengths of the sides of the kite, with equal sides as marked.

C7 Jill starts at High Point. She walks 12 km on a bearing of 030° to Much Higher and then 15 km on a bearing of 060° to Phew.

Work out how far (a) north (b) east of High Point Jill is.

C8 Calculate the volumes of the following solids.

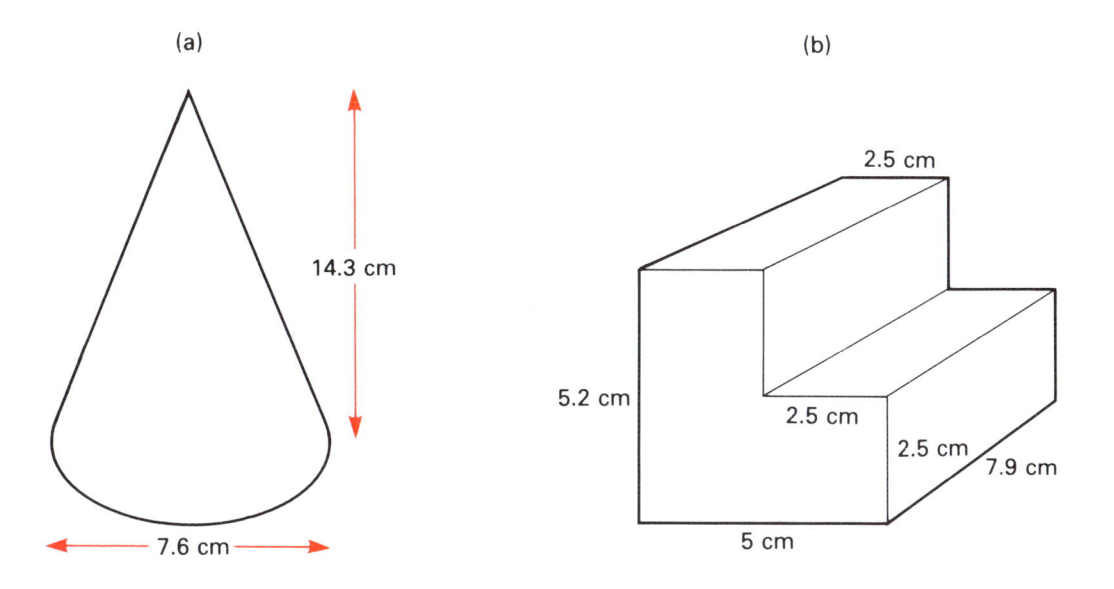

(a) 14.3 cm, 7.6 cm

(b) 2.5 cm, 5.2 cm, 2.5 cm, 2.5 cm, 7.9 cm, 5 cm

175

C9 Daffodils cost xp a bunch and tulips cost yp a bunch.

5 bunches of daffodils and 4 bunches of tulips cost £3·75.
8 bunches of daffodils and 5 bunches of tulips cost £5·30.

Work out the cost of a bunch of daffodils and a bunch of tulips.

D Mathematics for credit 1 review

D1 Write down (a) the next term rule and (b) nth term rule for the following sequences.

(i) 3, 7, 11, 15, . . .   (ii) 2, 8, 18, 32, . . .

D2 The area of a television screen varies as the square of its diagonal (d).

(a) What happens to the area of the screen if the diagonal's length is doubled?

(b) What happens to the length of the diagonal if the area is halved?

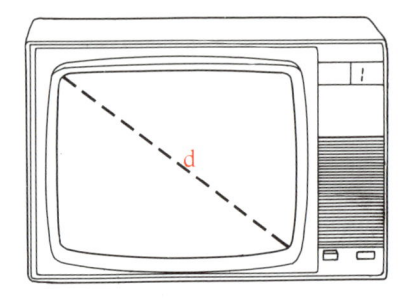

D3 Simplify the following surds.

(a) $\sqrt{45}$ (b) $\dfrac{2}{\sqrt{2}} + \tfrac{1}{2}$ (c) $\sqrt{2} + \sqrt{8} + 2\sqrt{12} - \sqrt{18}$

D4 Let $f(x) = 2x + 5$.

(a) Find $f(3)$ and $f(6)$.

(b) If $f(a) = 15$, find a.

D5 Solve the following for $0 \leqslant x \leqslant 360$.

Give your answers correct to 1 decimal place.

(a) $\sin x° = 0·7$ (b) $\cos x° + 1 = 0·203$

(c) $3 \tan x° + 4 = 7·632$

D6 A template for a piece of patchwork is to be made in the shape of a triangle with sides 10 cm, 8 cm and 5 cm.

Calculate the sizes of the angles of the triangle.

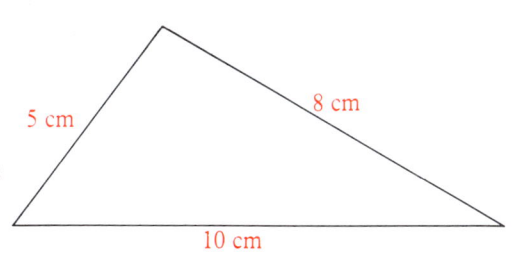

D7 Solve the following inequations.

(a) $3x - 15 \leq 27$ (b) $\dfrac{3x - 2}{4} > 7$

D8 Ann's annual salary of £14 750 is to be increased by 4·5%. Calculate her new monthly salary.

D9 Linsey throws a ball upwards from a balcony. The height of the ball (h m) above the ground is given by the formula
$h(t) = 5 + 4t - t^2$ where t is the time in seconds after she threw the ball.

(a) Draw the graph of $h(t)$ for $0 \leq t \leq 6$.

(b) When does the ball hit the ground?

(c) How high is the balcony above the ground?

(d) When does the ball pass the balcony again?

D10 The following instructions on how to estimate the distance across a river are given in the book *Scouting for Boys* by Lord Baden Powell. Explain clearly why this method works.

Distance Across a River

The way to estimate the distance across a river is to take an object X, such as a tree or rock on the opposite bank; start off at right angles to it from A, and pace, say, ninety yards along your bank; on arriving at sixty yards, plant a stick or stone, B; on arriving at C, thirty yards beyond that, that is ninety from the start, turn at right angles and walk inland, counting your paces until you bring the stick and the distant tree in line; the number of paces that you have taken from the bank C D will then give you the half distance across A X.

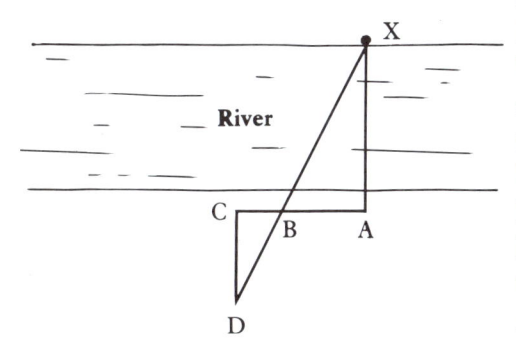

E Mathematics for credit 2 review

E1 Factorise the following.

(a) $x^6 - 16$ (b) $x^2 - 15x + 36$ (c) $3a^2 - 4ab + b^2$

E2 Simplify the following.

(a) $(4ab)^3$ (b) $\dfrac{7a^3 \times 4a^{\frac{2}{3}}}{2a}$ (c) $\dfrac{10c^4 \times 9c^{-2}}{3c^{-3}}$

E3 If a unit of electricity costs 5·2p, calculate the cost of running a 125 W stereo system for 9 hours.

E4 Calculate the circumference and area of this circle.

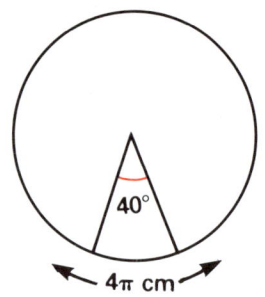

E5 Here is the pattern for the net of a box in the shape of a cuboid.

It has been draw on $\frac{1}{2}$ cm paper. To make the box the correct size, it must be redrawn on 2·5 cm squared paper.

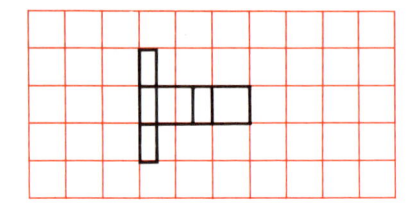

(a) What is the scale factor of the enlargement?

(b) What is the surface area of the real box?

(c) How many times bigger is the volume of the real box than that of the pattern?

E6 Sketch the graph of the function $f(x) = 2 \sin 2x° - 3$.

E7 Solve the following equations.

(a) $2^x = 256$ (b) $4^x = 8$ (c) $5^x = \frac{1}{625}$ (d) $8^x = 2$

E8 The sum of a number and its square is 72.

What might the number be?

Brainstormers

brainstormers [n]. Questions pertaining to reasoning and applications; stoaters; heidnippers, etc.

The questions in this section are taken from the SMP 11–16 pilot 16+ examination (papers 3 and 4) and are reprinted with the kind permission of the Oxford and Cambridge Schools Examination Board, and the East Anglian Examinations Board.

F1 A 'penny farthing' bicycle has wheels of two different sizes.

Their diameters are 125 cm and 39·5 cm.

(a) Calculate the circumference of

　(i) the large wheel　(ii) the small wheel

(b) How far does the cycle travel as the large wheel makes 20 revolutions? Give your answer in **metres**.

(c) How many revolutions does the small wheel make in travelling the same distance?

F2 A piston fits tightly in a vertical cylinder, trapping the air below it.

When a weight w kg is placed on the piston, the height of the piston above the base is h m.

h is inversely proportional to w.

(a) Sketch a graph of (w, h) on axes like these.

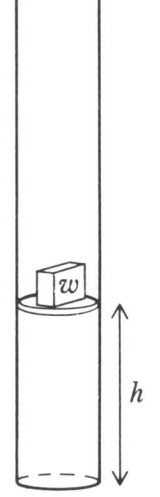

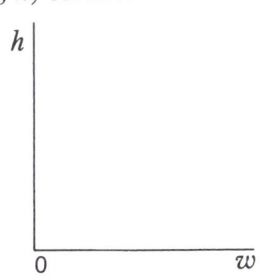

(b) What happens to h when w is halved?

(c) If $h = 45$ when $w = 250$, calculate h when $w = 300$.

F3 I travel to school by bus. Sometimes the bus arrives at the bus stop at the same time as I do. At other times I have to wait, for up to 10 minutes.

If the traffic lights are all green, the bus journey takes 7 minutes. Usually the journey takes longer, sometimes as long as 12 minutes.

It takes me 5 minutes to walk from home to the bus stop. It takes me 2 minutes to walk from the bus into school. I have to be in school by 8:45 a.m.

(a) By what time must I leave home to be **sure** of arriving at school on time?

(b) If I leave home at this time, what is the earliest I could arrive at school?

F4 (a) Write down the sum of the interior angles of a quadrilateral.

(b) Draw this pentagon.
Draw a diagonal in it.

Work out the sum of the interior angles of the pentagon.

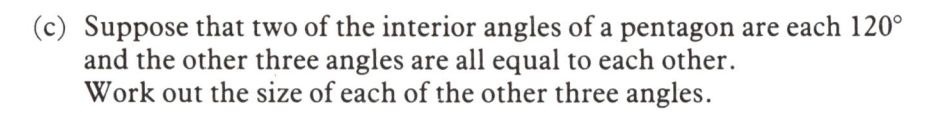

(c) Suppose that two of the interior angles of a pentagon are each 120° and the other three angles are all equal to each other.
Work out the size of each of the other three angles.

F5 The frequency of a note played on the E-string of a violin is **inversely** proportional to the length of the vibrating part of the string.

(a) What happens to the frequency of the note when the length of the vibrating part of the string is halved?

(b) Frequency is measured in hertz (Hz).
When the vibrating length is 204 mm, the frequency is 2048 Hz.
 (i) Calculate the frequency when the length is 250 mm.
 (ii) Calculate the length when the frequency is 3000 Hz.

F6 This diagram shows a cross-section of a symmetrical railway embankment. It is in the shape of a trapezium.
The diagram is not to scale.

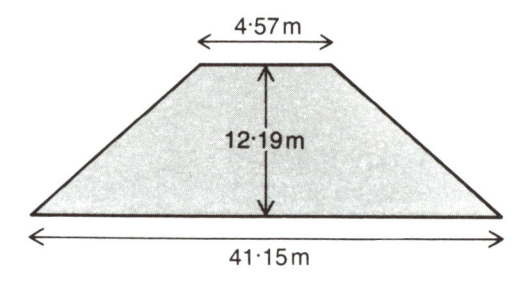

4·57 m
12·19 m
41·15 m

(a) (i) Calculate the area of the cross-section.

(ii) How many cubic metres of material are required to build a 100 m length of the embankment?

(b) This is a drawing of the same cross-section.

(i) Calculate the length AB.

(ii) Calculate the angle θ.

F7

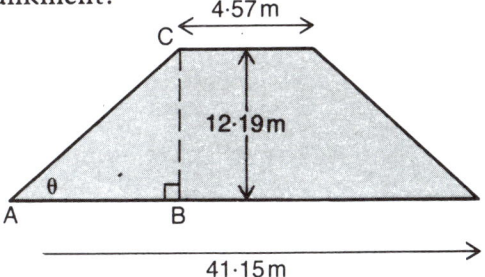

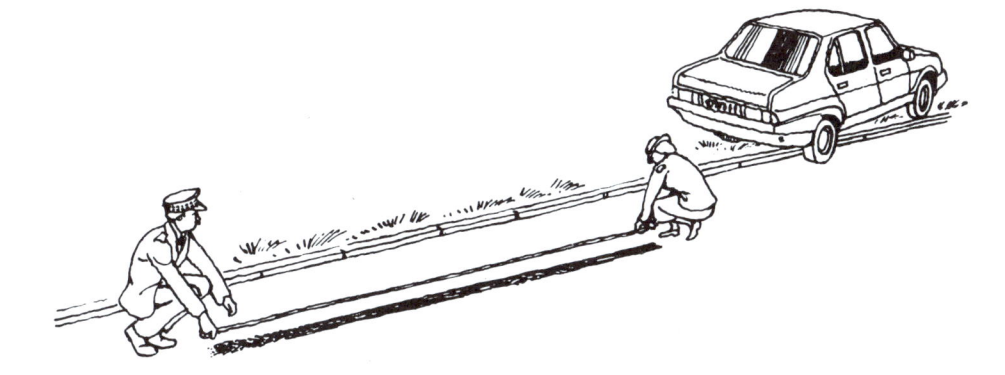

When measuring skid marks, the police can use this formula to estimate the speed of the vehicle.

$s = \sqrt{(30fd)}$

s is the speed in miles per hour (m.p.h.).
d is the length of the skid, in feet.
f is a number which depends on the weather and the type of road.

This table shows some values of f.

		Road surface	
		Concrete	Tar
Weather	Wet	0·4	0·5
	Dry	0·8	1·0

(a) A car travelling on a wet concrete road makes a skid mark of length 80 feet. How fast was it travelling?

(b) (i) When the road surface is tar and the weather is dry, the formula may be written

$s = \sqrt{(30d)}$

181

Complete this table to show the values of s for the given values of d, to 1 decimal place.

d	50	100	150	200	250
$30d$	1500				
$s = \sqrt{(30d)}$	38·7				

(ii) Draw axes, with d from 0 to 250 (use 2 cm for 50) and s from 0 to 100 (use 1 cm for 10).
Draw the graph of (d, s).

(iii) Use your graph to find how many feet a car would skid on a dry tar road at 75 m.p.h.

F8 In this question, you will need the formula

Surface area of disc = $2\pi r(r + t)$

where r is the radius and t the thickness.

The disease *multiple mycloma* causes the red blood cells, which are like discs, to stick together (like a pile of coins).

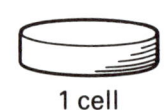

When the cells stick together, there are fewer faces to absorb oxygen.

1 cell

The cells have a thickness of 2·2 microns and a diameter of 7·2 microns.

4 cells stuck together

The surface area is measured in square microns.

(a) (i) What is the radius of a cell?

(ii) Find the surface area of one cell.

(iii) Find the total surface area of four separate cells.

(b) (i) Find the surface area of four cells stuck together.

(ii) Find the percentage decrease in surface area when four cells stick together.

F9 This table gives the first four terms of a sequence u.

u_1	u_2	u_3	u_4	...
3	6	12	24	...

(a) There is a simple relationship connecting each term with the next one. What is the relationship? Write it in words or in symbols.

(b) Write down a formula for the nth term u_n in terms of n.

F10 For cars travelling at normal speed, the air drag is roughly proportional to SV^2, where S is the surface area of the car and V is the speed.

By what number will the air drag be multiplied when the car increases speed from 40 km/h to 120 km/h?

F11 Opticians use the formula $D = \dfrac{1}{u} + \dfrac{1}{v}$.

What can you say about the value of D if $v = 0.02$ and u is approximately 1 000 000?

F12 A vacuum pump is designed to remove 25% of the gas in a vacuum chamber with each stroke.

(a) What percentage of the gas will be left after
(i) 1 stroke   (ii) 2 strokes

(b) How many strokes are needed to remove about 90% of the gas which was originally in the chamber?

F13 Roughly how many matchboxes of this size have a total volume of 1 cubic metre?

Show clearly how you get your estimate.

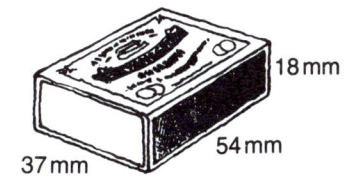

18 mm
54 mm
37 mm

F14 A sequence begins 1, 3, 7, 15, 31, 63, ···
It has a simple term-to-term rule.

(a) Write down this rule in words.

(b) Use the rule to find the next term of the sequence.

(c) The nth term of the sequence is denoted by s_n.
Write the rule as an equation connecting s_{n+1} and s_n.

(d) Make a new sequence by adding 1 to each term of the sequence above. Write an expression in terms of n for the nth term of the new sequence.

(e) Use your answer to part (d) to write down a formula for s_n in terms of n.

F15 x and y are given by these formulas.

$x = pt$ and $y = p(1 + qt)$

(a) Find a formula for y in terms of p, q and x which does not include t.

(b) Find a formula for x in terms of p, q and y which does not include t.

F16 The iteration formula of a sequence u is

$$u_{n+1} = \frac{u_n - 1}{4}$$

(a) Starting with $u_1 = 5$, calculate u_2, u_3, u_4, u_5 and u_6.

(b) Guess the value of the limit towards which the sequence seems to converge.

(c) Calculate the fixed point of the iteration formula (the value of u_1 for which $u_1 = u_2 = u_3$ etc.).

F17 A hospital keeps two types of glucose solution, weak and strong.

The weak solution contains 20 g of glucose per litre.
The strong solution contains 80 g of glucose per litre.

(a) Suppose x litres of the weak solution are mixed with y litres of the strong solution.

Write an expression, in terms of x and y, for

(i) the total volume of the mixture, in litres

(ii) the amount of glucose in the mixture, in grams

(b) A nurse needs to mix the weak and strong solutions to make 6 litres of a new solution, containing 45 g of glucose per litre.

Calculate the volume of weak solution and the volume of strong solution which she must mix together.

F18 Butter is to be supplied to Hillbury's Superstore in boxes containing 250 g packs of butter.

Each 250 g pack is a cuboid measuring 96 mm by 64 mm by 36 mm.

The internal dimensions of the boxes into which these are to be packed are 500 mm by 300 mm by 250 mm.

There are several different ways of placing one pack in the bottom corner of a box.

One way is like this.

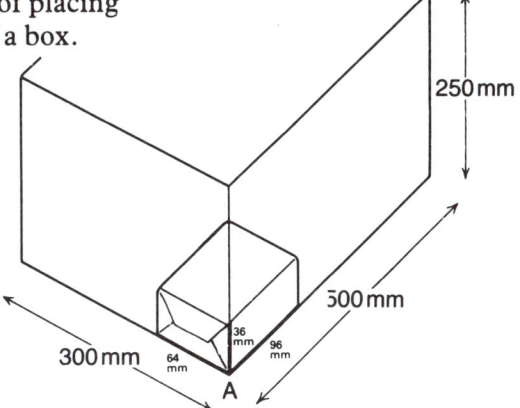

[Diagram showing box corner with dimensions: 250 mm height, 300 mm and 500 mm base edges, and a pack with dimensions 36 mm, 64 mm, 96 mm placed in corner A]

We can represent this way of placing the pack by this symbol:

[Symbol showing 36 up, 64 left, 96 right]

(a) Using similar symbols, list all the different ways of placing one pack in the bottom corner A of the box.

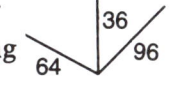

(b) When the packs of butter are put into the box they must all be the same way round. For each of the possibilities in (a), calculate how many packs can be fitted inside the box.

(c) (i) One of the solutions in part (b) gives the largest number of packs. Is it possible to make the box smaller and still contain this arrangement of packs? If so, what are the dimensions of the box which will just hold these packs?
   (ii) Is this a solution which the superstore is likely to adopt? Give a reason for your answer.

F19 The face of Brian's watch is decorated with two circles and a square.

The shaded part is gold.

One side of the square measures 20·0 mm.

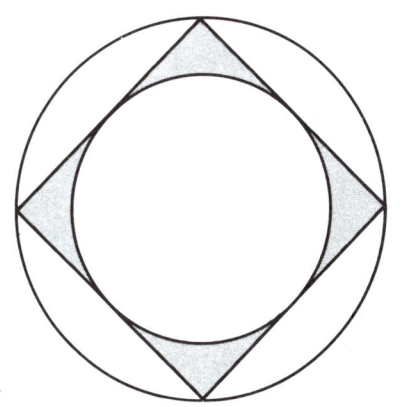

(a) What is the radius of the small circle?
(b) What is the area of gold?
(c) Calculate the radius of the large circle.

185

F20

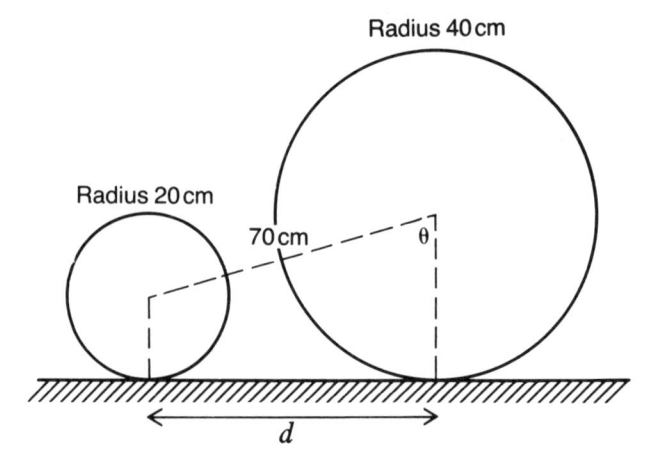

(a) This diagram shows two wheels, of radius 20 cm and 40 cm, standing on horizontal ground.

The distance between their centres is 70 cm.

(i) Calculate the distance marked d.

(ii) Calculate the angle marked θ.

(b) The same two wheels, with their centres still 70 cm apart, are connected by a tight belt, as shown below.

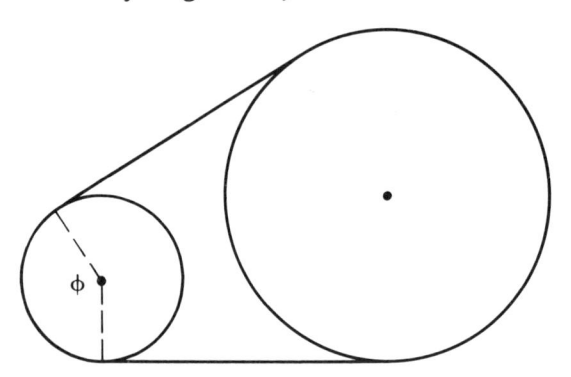

(i) Draw the diagram and add the line of symmetry.

(ii) Calculate the angle ϕ to the nearest degree.

(iii) What fraction of the circumference of the smaller wheel is touching the belt?

(iv) Calculate the total length of the belt, showing all your working.

F21 Pauline and Quentin have inherited this plot of land.

(a) Calculate the area of the plot.

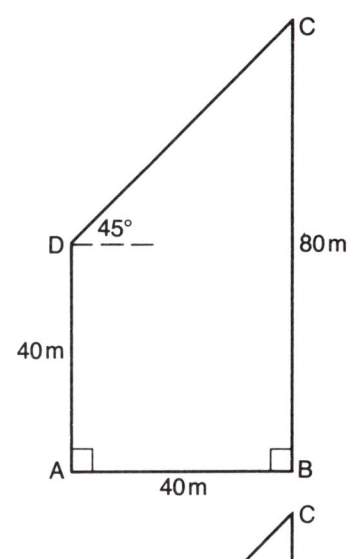

(b) (i) They agree to divide the plot into two equal parts by a straight fence parallel to AD and BC.

Suppose the fence is x metres from AD. Explain why the area to the left of the fence is $40x + \frac{1}{2}x^2$ square metres.

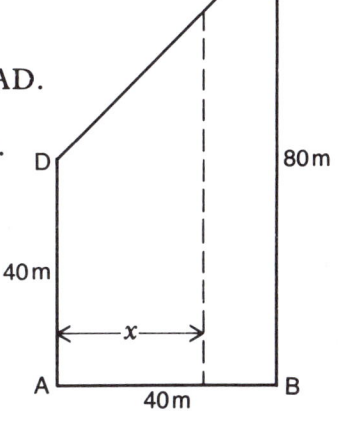

(ii) They want to choose x so as to divide the field as closely as possible into two equal parts. They can measure x to the nearest 0·5 m. Find by trial and error the value of x they should choose.

F22 Each shape below is made from a piece of wire of length 12 cm. Calculate the area of each shape, to the nearest 0·1 cm².

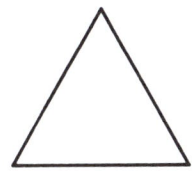

(a) Equilateral triangle

(b) Square

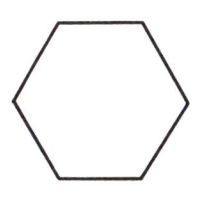

(c) Regular hexagon

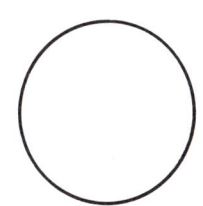

(d) Circle